MEMOIRS
of the
American Mathematical Society

Number 458

WITHDRAWN

Lyapunov Theorems
for Operator Algebras

Charles A. Akemann
Joel Anderson

November 1991 • Volume 94 • Number 458 (end of volume) • ISSN 0065-9266

American Mathematical Society
Providence, Rhode Island

1980 *Mathematics Subject Classification* (1985 *Revision*).
Primary 46L05, 46L30, 46L10; Secondary 28B05, 46B99,
46G10, 49E15, 52A20, 60A10, 93C15.

Library of Congress Cataloging-in-Publication Data

Akemann, Charles A.
 Lyapunov theorems for operator algebras/Charles A. Akemann, Joel Anderson.
 p. cm. – (Memoirs of the American Mathematical Society, ISSN 0065-9266; no. 458)
 Includes bibliographical references.
 ISBN 0-8218-2516-X
 1. Operator algebras. 2. Lyapunov functions. I. Anderson, Joel, 1939– . II. Title. III. Series.
QA3.A57 no. 458
[QA326]
510 s–dc20   91-28168
[512′.55]   CIP

Subscriptions and orders for publications of the American Mathematical Society should be addressed to American Mathematical Society, Box 1571, Annex Station, Providence, RI 02901-1571. *All orders must be accompanied by payment.* Other correspondence should be addressed to Box 6248, Providence, RI 02940-6248.

SUBSCRIPTION INFORMATION. The 1991 subscription begins with Number 438 and consists of six mailings, each containing one or more numbers. Subscription prices for 1991 are $270 list, $216 institutional member. A late charge of 10% of the subscription price will be imposed on orders received from nonmembers after January 1 of the subscription year. Subscribers outside the United States and India must pay a postage surcharge of $25; subscribers in India must pay a postage surcharge of $43. Expedited delivery to destinations in North America $30; elsewhere $82. Each number may be ordered separately; *please specify number* when ordering an individual number. For prices and titles of recently released numbers, see the New Publications sections of the NOTICES of the American Mathematical Society.

BACK NUMBER INFORMATION. For back issues see the AMS Catalogue of Publications.

MEMOIRS of the American Mathematical Society (ISSN 0065-9266) is published bimonthly (each volume consisting usually of more than one number) by the American Mathematical Society at 201 Charles Street, Providence, Rhode Island 02904-2213. Second Class postage paid at Providence, Rhode Island 02940-6248. Postmaster: Send address changes to Memoirs of the American Mathematical Society, American Mathematical Society, Box 6248, Providence, RI 02940-6248.

CONTENTS

ABSTRACT

In 1940 Lyapunov proved that the range of a nonatomic vector–valued measure is compact and convex. This theorem translates into the language of operator algebras as follows.

> Lyapunov's Theorem: If Ψ is a weak* continuous linear map from an abelian, nonatomic von Neumann algebra $\mathcal{M}$ to a finite dimensional space, then $\Psi(P) = \Psi((\mathcal{M}_+)_1)$, where P denotes the set of projections in $\mathcal{M}$ and $(\mathcal{M}_+)_1$ denotes the positive portion of the unit ball of $\mathcal{M}$.

In the present Memoir we take this theorem apart and perturb each hypothesis, thereby extending and generalizing Lyapunov's result in many directions. It is easy, for example, to delete the word "abelian" from the above theorem. Also, if N denotes the set of normal elements in the unit ball of $\mathcal{M}$ and U denotes the set of unitary elements in $\mathcal{M}$, then we show that $\Psi(N) = \Psi(U)$. When $\mathcal{M}$ is atomic, we prove an approximate version of Lyapunov's theorem. When $\mathcal{M}$ is countably decomposable with finite dimensional finite summand and Ψ is singular, then Lyapunov's theorem holds even when the range of Ψ is infinite dimensional. In the first section of this Memoir we prove an abstract version of Lyapunov's theorem in which Ψ is a continuous affine map of a compact convex subset of a locally convex space into a finite dimensional space. In section 7 we specialize to the "hardest" case, where $\Phi : \mathcal{M} \rightarrow \mathcal{N}$ is a positive linear map, $\mathcal{M}$ is atomic and abelian and $\mathcal{N}$ is a full matrix algebra. Only an approximate Lyapunov type theorem is possible in this case, and the difficulties are caused by the choice of the operator norm on $\mathcal{N}$. Results in section 7 and 8 are related to the Kadison-Singer problem of uniqueness of extension of pure states from atomic maximal abelian subalgebras of B(H).

Received by the editor January 11, 1991.

§0. INTRODUCTION AND NOTATION

In 1940 A. A. Lyapunov published his celebrated discovery of the fact that
the range of a nonatomic vector–valued measure is convex and compact [33].
Since that time this beautiful theorem has intrigued and inspired scholars in
many disciplines including combinatorics [41], control theory [36,44], economics
[see 44], functional analysis [9,15,18,23,24,29,31,46,47], graph theory [10], logic
[7,32], probability [17,30] and topology [16]. The purpose of the present work is
to present new generalizations of Lyapunov's theorem and related results, mainly
in an operator algebraic setting.

We were led to this study while engaged in an attempt to solve a well–known
problem of Kadison and Singer concerning the uniqueness of pure state extensions of complex homomorphisms on atomic maximal abelian subalgebras of
$B(H)$. We found that this problem could be recast as a conjecture that is a
natural generalization of Lyapunov's theorem. A bit of progress was made to-
wards proving the conjecture (see section 7) and, while engaged in this work, we
discovered several other generalizations of Lyapunov's theorem. Thus, we were
led to a systematic examination of Lyapunov theorems for operator algebras.

Our point of view follows that of Lindenstrauss [31]. So, in our terminology,
Lyapunov's theorem asserts that if Ψ is a weak* continuous linear map of an
abelian, nonatomic von Neumann algebra $\mathcal{N}$ to a complex vector space of finite
dimension, and $a \in (\mathcal{N}_+)_1$, the positive part of the unit ball of $\mathcal{N}$, then there is
an extreme point p of $(\mathcal{N}_+)_1$ (i.e., a projection) such that $\Psi(p) = \Psi(a)$.

Given this perspective, the most natural setting for a study of more general
results is a linear space $\mathcal{X}$ containing a convex set Q and an affine map Ψ of Q
into a linear space $\mathcal{Y}$. By a **Lyapunov theorem** we mean a result that gives
conditions which ensure that, for each x in Q, there is an extreme point e of Q
such that $\Psi(e)$ is "close" to $\Psi(x)$.

Of course, the most satisfying conclusion would be that $\Psi(E(Q)) = \Psi(Q)$,
where $E(Q)$ denotes the extreme points of Q. We call such a theorem a **Lyapunov theorem of type 1**. If we assume that $\mathcal{Y}$ is a normed linear space so
that the notion of closeness can be made precise, then there are 3 other natural
types of Lyapunov theorems:

> **Type 2 concludes that $\Psi(E)$ is norm dense in $\Psi(Q)$.**
> **Type 3 concludes that $\Psi(E)$ is ϵ–dense in $\Psi(Q)$.**
> **Type 4 concludes that Ψ is not isometric on E, when Ψ has
> norm one.[1]**

---

[1]Of course, we must assume that $\mathcal{X}$ is a normed space in this case

1

Results of all 4 types will be presented here. In most cases, the range of Ψ will be assumed to have finite dimension and often it will be necessary to impose conditions that ensure that Q is compact. In some cases (notably in §6) it is possible to get results in the case where the range of Ψ is infinite–dimensional. Also, in some cases Q need not be compact (§3 and §6).

Before discussing our results in more detail, it is useful to fix some of the notation that will be used throughout the paper. We shall use Ψ to denote an affine map defined on a convex subset of linear space. The symbols $\mathcal{M}$ and A shall always stand for a von Neumann algebra and a C^*–algebra, respectively. If S is a subset of a normed linear space $\mathcal{X}$, then we write

$$(S)_1 = \{x \in S : \|x\| \leq 1\}.$$

Also, we let

$$A_{sa} \text{ and } A_+$$

denote the self–adjoint elements and the positive elements in the C^*–algebra A.

We shall derive Lyapunov theorems for operator algebras where the convex set Q is taken to be $(A_+)_1$, $(A_{sa})_1$ or $(A)_1$ (or faces thereof). We write $(A_\#)_1$ (or $(\mathcal{M}_\#)_1$ in the weakly closed case) to indicate one of these sets. The structure of the extreme points of $(A_\#)_1$ has been known for some time. We now review this information. The extreme points of $(A_+)_1$ are the projections in A [27,7.4.6]; we denote this set by P_A (or just by P, when the ambiant algebra is obvious). Recall that a **symmetry** is an element s in A such that $s = s^*$ and $s^2 = 1$. We denote the set of symmetries in A by S_A (or simply S). As is well–known (and easy to see), an element s of A is a symmetry if and only if $s = 2p - 1$, where p is a projection. Since the map $a \mapsto 2a - 1$ is an affine isomorphism from $(A_+)_1$ to $(A_{sa})_1$ (when A is unital), it is immediate that the extreme points of $(A_{sa})_1$ are the symmetries S. If p and q are projections in A, then [40,1.6.5] the extreme points of $(pAq)_1$ are precisely the elements x in $(pAq)_1$ such that

$$(p - xx^*)A(q - x^*x) = \{0\}.$$

Moreover, if x is an extreme point in $(pAq)_1$, then x is a partial isometry. In particular, if A is unital, then the set U of unitaries consists of extreme points in $(A)_1$.

The range of Ψ will often be taken to be the compex vector space $\mathbb{C}^n$, where n is finite. Given a convex set Q in a linear space, its extreme points are denoted by $E(Q)$. We write $B(H)$ for the bounded linear operators acting on a complex, separable, Hilbert space H. Usually, we take H to have infinite dimension; occasionally, however, (in §7, for example) H may be chosen to be finite–dimensional. Finally, We say $\mathcal{N}$ is a MASA in $\mathcal{M}$ if $\mathcal{N}$ is a maximal abelian C^*–subalgebra of $\mathcal{M}$.

Let $\mathcal{M}$ denote a von Neumann algebra. Since the family of minimal projections in $\mathcal{M}$ is unitarily invariant, there is a central projection z_a in $\mathcal{M}$ that is the supremum of all the minimal projections in $\mathcal{M}$. We say that $\mathcal{M}$ is **atomic** if $z_a = 1$ and $\mathcal{M}$ is **nonatomic** if $z_a = 0$.

If we put $z_{na} = 1 - z_a$ and write

$$\mathcal{M}_a = z_a \mathcal{M} \quad \text{and} \quad \mathcal{M}_{na} = z_{na} \mathcal{M},$$

then we have the decomposition

$$\mathcal{M} = \mathcal{M}_a \oplus \mathcal{M}_{na}$$

into its atomic and nonatomic components. Write $z_{fin} = \sup\{z \in \mathcal{M}_a : z \text{ is a}$ central projection and $z\mathcal{M}_a$ is finite dimensional$\}$ and put $z_{inf} = z_a - z_{fin}$. If we write

$$\mathcal{M}_{fin} = z_{fin} \mathcal{M} \quad \text{and} \quad \mathcal{M}_{inf} = z_{inf} \mathcal{M},$$

then we have that $\mathcal{M}_{fin}$ is the direct sum of finite type I factors and $\mathcal{M}_{inf}$ is the direct sum of infinite type I factors. Moreover, we have the decomposition

$$(*) \qquad \qquad \mathcal{M} = \mathcal{M}_{na} \oplus \mathcal{M}_{inf} \oplus \mathcal{M}_{fin}.$$

Note that $\mathcal{M}_{fin}$ is just the finite part of the atomic algebra $\mathcal{M}_a$. Of course, $\mathcal{M}_{na}$ may contain a finite (type II) summand. The decomposition $(*)$ above is appropriate here because of the nature of our results. In fact we shall derive a Lyapunov theorem of type 1 for $\mathcal{M}_{na}$ (Theorem 2.5), a Lyapunov theorem of type 2 for $\mathcal{M}_{inf}$ (Theorem 3.12) and a Lyapunov theorem of type 3 for $\mathcal{M}_{fin}$ (Theorem 4.1).

We now give a brief description of our results.

In section 1 we prove a result that we call an abstract Lyapunov theorem. Before presenting it, some elementary facts from convexity theory are discussed. The notion of "facial dimension" of a convex set Q is introduced here. Roughly speaking, this number is the infimum of the dimension of the "flat spots" on the boundary of Q. We also give conditions which insure that a (continuous) affine map on Q can be extended to a (continuous) linear map on the real space generated by Q (Proposition 1.5). In Theorem 1.6 we show that if Ψ is an affine map on Q and the dimension of of the span of $\Psi(Q)$ is less than the facial dimension of Q, then each extreme point of $\Psi^{-1}(x) \cap Q$ is an extreme point of Q for each x in Q. Using this we get a Lyapunov theorem of type 1 (Corollary 1.7) when $\mathcal{X}$ is locally convex, Ψ is continuous with finite dimensional range, Q is compact and the facial dimension of Q is larger than the dimension of the range.

We also show that a theorem of type 2 holds for Ψ^* if Ψ is a compact map into an infinite–dimensional normed space. We then give examples that show that it is not enough to assume that Ψ is weakly compact and that, in general, we cannot expect to get a type 1 result even when Ψ is compact. The section ends with 2 applications. A generalized Phelps–Dye theorem is derived and a more transparent route to a theorem of Glicksberg on Birmbaum–Orlicz spaces is indicated.

In section 2 we study nonatomic von Neumann algebras. In this case, it is possible to derive results when the subset Q is a weak* closed face of $(\mathcal{M}_{\#})_1$ or $(p\mathcal{M}q)_1$ (where p and q are projections). It is assumed that Ψ is an affine, weak* continuous map of Q and that the range is finite–dimensional. Theorem

2.5 asserts that a type 1 Lyapunov theorem holds in this situation. We show in Theorem 2.2 that if Q is a face as above, then it is affinely isomorphic to a set of the form $(\mathcal{N}_\#)_1$ or $(p\mathcal{N}q)_1$ for a nonatomic subalgbra $\mathcal{N}$. Thus, the proof of 2.5 reduces to the case where Q is just $(\mathcal{M}_\#)_1$ or $(p\mathcal{M}q)_1$. Since the facial dimension of such sets is infinite (Lemmas 2.3 and 2.4), Theorem 2.5 follows from Corollary 1.7. Given Theorem 2.5, Lyapunov's original result is easy to derive (Corollary 2.6). Finally, a simple example is given which shows that if we allow the range to be infinite dimensional, then not even a type 4 result is possible without more assumptions.

In section 3 we consider C^*–algebras. We first establish a density theorem (Theorem 3.2) that characterizes the weak* closures of the sets of projections and unitaries in a von Neumann algebra. Using this, we show that, if A is a unital C^*–algebra with no finite-dimensional representations (in which case we say that A is **weakly infinite**), then the set U_A of unitaries is weakly dense in $(A)_1$ (Theorem 3.4). Moreover, if A has sufficiently many projections, then we have that P_A is weakly dense in $(A_+)_1$ (Theorem 3.5). It is then easy to show that, if A is weakly infinite and Ψ is a norm continuous map of A into $\mathbb{C}^n$, then a Lyapunov theorem of type 2 holds for $Q = (A)_1$. Also, if A also has sufficiently many projections, then a Lyapunov theorem of type 2 holds for $Q = (A_+)_1$ (Theorem 3.6).

Next, examples are given to show that a type 1 result need not hold for weakly infinite C^*–algebras and that no type 2 theorem is possible in the non–weakly infinite case (e.g., when A is the algebra of 2×2 complex matrices). Finally, we present a Lyapunov theorem for weakly infinite von Neumann algebras (Theorem 3.12). In this case it is automatic that there are sufficiently many projections (and symmetries). Thus, the case where $Q = (\mathcal{M}_\#)_1$ follows as an immediate corollary from Theorem 3.6. In addition it is shown that we may also take $Q = (p\mathcal{M}q)_1$.

Finite atomic von Neumann algebras are considered in Section 4. Since such an algebra is the direct sum of finite matrix algebras, the example in Section 3 mentioned above shows that the best one can hope for in this case is a type 3 Lyapunov theorem. Such a result is derived in Theorem 4.1. It is assumed that the map Ψ is weak* continuous and its range is $\mathbb{C}^n$ with the ℓ^∞ norm. The convex set Q may be taken to be $(\mathcal{M}_\#)_1$ or $(p\mathcal{M}q)_1$. It is shown that the range of the extreme points of Q is δ–dense (or in some cases 2δ–dense) in $\Psi(Q)$. The constant δ is defined by the formula

$$\delta = \sup\{\|\Psi(p)\|_1 : p \text{ is a minimal projection in } \mathcal{M}\}.$$

Since infinite atomic von Neumann algebras are automatically weakly infinite, theorems 2.5, 3.12 and 4.1 can be combined to get a Lyapunov theorem for arbitrary von Neumann algebras. This is recorded in Corollary 4.3. Next we apply Theorem 4.1 to the case where $\mathcal{M} = \ell^\infty(m)$ (m finite) and Ψ is determined by an $n \times m$ matrix (Corollary 4.4). It is shown that if the ℓ^1–norm of the columns of Ψ are no greater than γ, then the image of the extreme points of $(\ell^\infty(m)_\#)_1$ are γ–dense in $\Psi((\ell^\infty(m)_\#)_1)$. In particular we get that if each row of Ψ has

ℓ^1–norm equal to 1, then it is possible to find a subset $\sigma \subset \{1, \ldots, n\}$ such that

$$\left| \sum_{i \in \sigma} |\omega_{ij}| - \frac{1}{2} \right| < \gamma, \qquad j = 1, \ldots, n,$$

where ω_{ij} denotes the (i, j)–entry of Ψ. (See remark 2 following Theorem 7.12 for further details). We next use Corollary 4.4 get a combinatorial theorem first proved by Beck and Fiala [10].

Finally, after presenting some examples, it is shown that a type 3 Lyapunov theorem holds when $\mathcal{M}$ is an arbitrary von Neumann algebra and the range of Ψ lies in $\ell^\infty(\mathbb{N})$. Here Ψ must be weak* continuous; the conclusions are analogous to those of Theorem 4.1. (This result does not contradict the example of section 2 mentioned above because, in that case, we have $\delta = \infty$).

We study simultaneous approximations in §5. We only consider abelian von Neumann algebras in this section. The nonatomic case is treated in Theorem 5.3. It is shown that if $\{a_1, \ldots, a_m\}$ are positive elements in $\mathcal{M}$ which sum to 1, Ψ is weak* continuous and maps to $\mathbb{C}^n$, then there are projections $\{p_1, \ldots, p_m\}$ which also sum to 1 and such that $\Psi(p_i) = \Psi(a_i), i = 1, \ldots, m$. In the atomic case, which is presented in Theorem 5.7, we must assume each $a_i = \frac{1}{m}$; moreover, in this case, we can only get a density result, since the proof relies on Theorem 4.1.

In Section 6 singular maps of von Neumann algebras are considered. For such maps, the domain has an "essentially" inseparable character. Moreover, the proofs of the results in the previous sections hinged, in good part, on the fact that the dimension of the domain was strictly greater than that of the range. Thus, given the "inseparable" nature of the domain in the singular case, one might expect that the range can now be allowed to be infinite–dimensional. In fact this turns out to be the case. In Theorem 6.12, it is shown for many von Neumann algebras $\mathcal{M}$ that if Ψ is a singular, norm continuous map of $\mathcal{M}$ into a normed linear space whose dual is weak* separable, then a Lyapunov theorem of type 1 holds with conclusions analogous to those in Theorem 2.5 (although weak* closed faces are treated separately). For this theorem to hold, the following assumptions are necessary. First, our technique does not work when the center of the finite part of $\mathcal{M}$ (including the finite part of $\mathcal{M}_{na}$) has infinite dimension and so we must assume that this part of the center has finite dimension. Second, certain set theoretic difficulties were encountered in the case where $\mathcal{M}$ is not countably decomposable. It was therefore necessary to also impose a restriction on the size of $\mathcal{M}$. We note, however, that Theorem 6.12 certainly holds if $\mathcal{M}$ is a countably decomposable factor.

Since the required set theory is somewhat technical, the section begins with a review of the salient facts. In Definition 6.3 we state what it means for a von Neumann algebra $\mathcal{M}$ to be **essentially countably decomposable**. It is then shown in Propsition 6.4 that $\mathcal{M}$ is essentially countably decomposable if it is countably decomposable or else if the continuum hypothesis is true and every family of orthogonal projections in $\mathcal{M}$ has cardinality less than that of the first measureable cardinal. It should be noted that it is unlikely that a working operator algebraist (willing to assume the continuum hypothesis) would ever encounter a von Neumann algebra which was not essentially countably decomposible. After some additional set theoretic and operator algebraic preliminaries, Theorem 6.12 is presented. It is shown in Corollary 6.13 that the conclusions of Theorem 6.12 also hold in the case where Q is a weak* closed face of $(\mathcal{M}_+)_1$.

After presenting some examples, we derive two applications of Theorem 6.12. In particular, it is shown in Corollary 6.19 that the conclusions of Theorem 6.12 hold when Ψ is a conditional expectation of $B(H)$ (H separable) onto an injective nonatomic von Neumann algebra. The final part of this section is devoted to the proof of a "combination" Lyapunov theorem. If Ψ is a norm continuous map on a von Neumann algebra $\mathcal{M}$ into $\mathbb{C}^n$, then we have the decomposition $\Psi = \Psi_n + \Psi_s$, where Ψ_n is normal and Ψ_s is singular. In Theorem 6.23 we combine the results of Theorems 2.5 and 6.12 to show that a type 1 Lyapunov theorem holds for norm continuous maps of certain nonatomic von Neumann algebras into $\mathbb{C}^n$. Here Q is $(\mathcal{M}_+)_1$ or a weak* closed face thereof. In order to apply Theorem 6.12 we must assume that $\mathcal{M}$ is essentially countably decomposible and that the center of the finite part of $r\mathcal{M}r$ is small for each nonzero projection r. Finally, we show that that one can easily obtain a variant of a theorem of Choda, et al. [15] from Theorem 6.23.

In Section 7 we turn our attention to connections with the problem of Kadison and Singer mentioned on page 1. We consider the map Φ of an atomic maximal abelian subalgebra (i.e., a MASA) of $B(H)$ into $B(H)$ given by $\Phi(x) = pxp$, where p is a projection in $B(H)$. Note that the range of Φ is noncommutative so that the ℓ^∞ norm on the range is now replaced by the operator norm. We present several related conjectures concerning Φ. These conjectures are the natural analogs of the type 3 Lyapunov result proved in Theorem 4.1. We show how the truth of the conjectures implies a positive answer to the question raised by Kadison and Singer and then show that a Lyapunov theorem of type 4 holds for Φ when p has finite rank. It is interesting to note that although this theorem has an analytic conclusion, it is proven using a result about matroids, a notion from combinatorics. We conclude this section by presenting two examples.

In the final section of the paper applications to the paving problem are presented.

Throughout the paper we use the symbol $\square$ to denote the end of a proof. The table given on the following page summarizes our main results.

Summary of the Main Results

Ref.	Domain	Q	Range	Ψ	Additional Conditions	Type
1.7	L.C.S., $\mathcal{X}$	compact	$\mathbb{C}^n$	contin.	facial dim. $> n$	1
2.5	Nonatomic v.N. alg., $\mathcal{M}$	faces of $(\mathcal{M}_\#)_1$	$\mathbb{C}^n$	weak* contin.		1
3.6	C^*-alg., A	$(A_\#)_1$	$\mathbb{C}^n$	contin.	A, weakly infinite	2
3.12	v.N. alg., $\mathcal{M}$	$(\mathcal{M}_\#)_1$	$\mathbb{C}^n$	contin	$\mathcal{M}$, weakly infinite	2
4.1	Finite, atomic v.N. alg, $\mathcal{M}$	$(\mathcal{M}_\#)_1$	$\mathbb{C}^n$	weak* contin.		3
4.3	v.N. alg., $\mathcal{M}$	$(\mathcal{M}_\#)_1$	$\mathbb{C}^n$	weak* contin.		3
4.8	v.N. alg., $\mathcal{M}$	$(\mathcal{M}_\#)_1$	$\ell^\infty(\mathbb{N})$	weak* contin.		3
6.12	v.N. alg., $\mathcal{M}$	$(\mathcal{M}_\#)_1$	Normed space $\mathcal{X}$	contin., singular	$\mathcal{M}$, ess. count. decomp.; center of finite part of $\mathcal{M}$ finite dimensional; $\mathcal{X}^*$, weak*separable	1
6.23	Nonatomic v.N. alg., $\mathcal{M}$	face of $(\mathcal{M}_+)_1$	$\mathbb{C}^n$	contin., affine	Same as 6.12 and $\mathcal{M}$ is nonatomic	1
7.12	Diagonal MASA, $\mathcal{D}$	$(\mathcal{D}_{sa})_1$	$B(H)$	$\Phi(x) = pxp$	p, finite rank projection	4

§1. AN ABSTRACT LYAPUNOV THEOREM

In this section we study the question of finding the most general setting in which a Lyapunov theorem holds. Since the conclusion of such a theorem will concern extreme points of the convex set Q, we should at least work in the category of linear spaces with morphisms that preserve convexity. Thus, we will assume that the map Ψ is an affine map defined on Q and taking values in a linear space. Surprisingly, it is possible, even in this minimal environment, to derive a result that deserves to be called a Lyapunov theorem (Theorem 1.6). Of course, when working in this generality, one cannot expect to be guaranteed the existence of extreme points, and so our theorem only concludes that, if $v \in Q$, then every extreme point of $\Psi^{-1}(\Psi(v))$ is an extreme point of Q. In order to get this conclusion, we must assume that the boundary of Q does not contain "flat spots" with dimension smaller than the dimension of the range. If we impose the additional conditions that Q be a compact subset of a locally convex space (so that, by the Krien-Milman theorem, Q contains an abundance of extreme points) and that the map Ψ be continuous, then it follows from our abstract Lyapunov theorem that

$$\Psi(E(Q)) = \Psi(Q).$$

It is convenient to begin by reviewing some elementary ideas from convexity theory and introducing some notation. Most of the following results are surely known to experts in convexity theory. However, they do not seem to be available in any reference that is familiar to operator algebraists; so this material will be treated in some detail. Throughout this section Q **shall denote a convex subset of a** (real) **linear space** $\mathfrak{X}$. If x and y are in Q, then the line segment joining x and y is denoted by $\operatorname{seg}(x,y)$. Thus,

$$\operatorname{seg}(x,y) = \{\, \lambda x + (1-\lambda)y : \lambda \in [0,1] \,\}.$$

If x and y are in Q and $x \neq y$, then an element $v \in \operatorname{seg}(x,y)$ is said to be an **internal point** of $\operatorname{seg}(x,y)$ if $v \neq x$ and $v \neq y$. We adopt the convention that v is an internal point of $\operatorname{seg}(v,v)$. Recall that a convex subset F of Q is called a **face** of Q if, given $v \in F$ and $x,y \in Q$ such that v is an internal point of the $\operatorname{seg}(x,y)$, then $\operatorname{seg}(x,y) \subset F$. Thus, in this notation, a point b in Q is an **extreme point** of Q if $\{b\}$ is a face of Q. If v is in Q but is not an extreme point, then v is said to be a **composite point** of Q.

If v and y are in Q, then v is said to be **internal relative to y** if there exists $\alpha > 0$ such that $(1+\alpha)v - \alpha y$ is in Q and v is said to be a **weak internal point** of Q if v is internal relative to y for every y in Q. Given $v \in Q$ we write

$$G(Q,v) = \{\, y \in Q : v \text{ is internal relative to } y \,\}.$$

Note that v is always internal relative to itself so that $v \in G(Q, v)$.

The following proposition clarifies the relationships among these concepts.

PROPOSITION 1.1.

(1) If v and y are in Q and $y \neq v$, then v is an internal point relative to y if and only if there is an x in Q such that v is an internal point of $\text{seg}(x, y)$.
(2) If v is in Q, then v is a weak internal point of Q if and only if 0 is an weak internal point of $Q - v$.
(3) $G(Q, v)$ is the union of all line segments in Q that contain v as an internal point.
(4) $G(Q, v) = \{v\}$ if and only if v is an extreme point of Q.

PROOF. To show that (1) holds, note that, by definition v is internal relative to y if and only if there is $\alpha > 0$ such that, with $x = (1 + \alpha)v - \alpha y$, we have $x \in Q$. If this is the case, then we may write

$$v = \left(\frac{1}{1 + \alpha} \right) x + \left(\frac{\alpha}{1 + \alpha} \right) y$$

and get that $v \in \text{seg}(x, y)$. Conversely, if $v = \lambda x + (1 - \lambda)y$ for some $x \in Q$, then v is internal relative to y (with $\alpha = \frac{1}{\lambda} - 1$). So, (1) is true.

If v is a weak internal point of Q and $y \in Q$, then by part (1) there is an x in Q such that v is an internal point of the line segment joining x and y. In this case 0 is an internal point of $\text{seg}(x - v, y - v)$, and so 0 is an internal point of $Q - v$. The reverse implication follows by a similar arguement. Hence (2) holds. Statement (3) is also an easy consequence of (1), and (4) follows from (3). $\square$

The content of our next result, which is a bit more substantial, is that $G(Q, v)$ is the smallest face that contains v and v is a weak internal point of this convex set. It should be noted that these ideas are related to the standard notion of the relative interior of a convex set in $\mathbb{R}^n$ [38,Chapter 6]. In that setting, v is in the relative interior of Q if and only if $G(Q, v) = Q$. See the paragraph preceding 6.20 for more details.

THEOREM 1.2.

(1) For each $v \in Q$, $G(Q, v)$ is a face of Q.
(2) For each $v \in Q$, v is a weak internal point of $G(Q, v)$.
(3) If H is a face of Q that contains v, then $G(Q, v) \subset H$.
(4) If H is a face of Q and v is a weak internal point of H, then $H = G(Q, v)$.

PROOF. To simplify the notation, write $G = G(Q, v)$. We begin by showing that G is convex. Fix y in G and select $\alpha > 0$ such that $(1 + \alpha)v - \alpha y$ is in Q. For any $0 < \beta < \alpha$ we have

$$\frac{\beta}{\alpha}[(1 + \alpha)v - \alpha y] + \left(1 - \frac{\beta}{\alpha} \right) v = (1 + \beta)v - \beta y,$$

and so $(1 + \beta)v - \beta y$ is in Q. Thus, if x and y are in G, we may find a positive α such that both $(1 + \alpha)v - \alpha x$ and $(1 + \alpha)v - \alpha y$ are in Q. If we fix λ in $(0, 1)$, then a simple calculation shows that

$$(1 + \alpha)v - \alpha[\lambda x + (1 - \lambda)y] = \lambda[(1 + \alpha)v - \alpha x] + (1 - \lambda)[(1 + \alpha)v - \alpha y],$$

and the right hand side lies in Q by convexity. Hence, $\lambda x + (1 - \lambda)y$ is in G and therefore G is convex.

To show that G is a face, by a standard argument, it suffices to show that if G contains the midpoint of a line segment in Q, then the endpoints of the line segment also lie in G. So, suppose x and y are in Q and $w = (x + y)/2$ is in G. Pick $\alpha > 0$ such that $(1 + \alpha)v - \alpha w$ is in Q and set

$$\lambda = \frac{2}{2 + \alpha}.$$

It is easy to check that

$$\lambda[(1 + \alpha)v - \alpha w] + (1 - \lambda)x = \left(1 + \frac{\alpha}{2 + \alpha}\right)v - \left(\frac{\alpha}{2 + \alpha}\right)y,$$

and therefore $y \in G$. Repeating this argument with x and y interchanged, we get that x also lies in G, and so G is a face of Q. Part (2) is now an immediate consequnce of the definition of G.

To see that (3) holds, let H denote a face of Q that contains v, and fix y in G. If we pick $\alpha > 0$ so that $(1 + \alpha)v - \alpha y$ is in Q, then we have

$$\left(\frac{1}{1 + \alpha}\right)((1 - \alpha)v - \alpha y) + \left(1 - \frac{1}{1 + \alpha}\right)y = v.$$

Since H is a face, y is in H, and so $G \subset H$.

Finally, if H is a face of Q such that v is a weak internal point of H, then for each y in H, v is an internal point relative to y, and so $H \subset G$. Since we have $G \subset H$ by part (3), $G = H$ and the proof is complete. $\square$

Next we introduce the notion of facial dimension, $\dim_{\mathcal{F}}(\cdot)$, for a convex set. For each $v \in Q$ we let $\mathbf{dim}(Q, v)$ denote the dimension of the real linear span of $G(Q, v) - v$.[1] Note that v is an extreme point of Q if and only if $\dim(Q, v) = 0$. We define the **facial dimension** of Q by

$$\dim_{\mathcal{F}}(Q) = \min\{\dim(Q, v) : v \text{ is a composite point of } Q\}.$$

(If Q does not contain a composite point, then Q consists of a single point. We do not define $\dim_{\mathcal{F}}(Q)$ in this trivial case). If H is a face of Q that contains the composite point v, then by part (3) of Theorem 1.2, $G(Q, v) \subset H$, and so the dimension of the linear span of $H - v$ is at least $\dim_{\mathcal{F}}(Q)$. Thus, $\dim_{\mathcal{F}}(Q)$

---

[1] In the parlance of convexity theory, $\dim(Q, v)$ stands for the dimension of $G(Q, v)$. See pages 63–4 for further details.

measures the minimal "size" of the faces of Q that contain composite points. Note that, since Q itself is a face in Q, we have that $\dim_{\mathcal{F}}(Q) \geq 1$ (if Q contains a composite point v). In particular, if Q is smooth in the sense that every other face in Q is an extreme point, we have that $\dim_{\mathcal{F}}(Q)$ is the linear dimension of the real subspace spanned by $Q - v$.

If S is a subset of $\mathcal{X}$, then we denote the real subspace that S generates by $\mathbb{R}$-**span**(S). Also, S is said to be **radial** if for every $x \in \mathcal{X}$ there is $\alpha > 0$ such that $\pm \alpha x \in \mathcal{X}$.

PROPOSITION 1.3. *If 0 is a weak internal point of Q and $\mathbb{R}$–span$(Q) = \mathcal{X}$, then Q is radial.*

PROOF. Fix x in $\mathcal{X}$. Since $\mathbb{R}$–span$(Q) = \mathcal{X}$, there is a subset $\{x_1, \ldots, x_n\}$ in Q and nonzero real scalars $\{\lambda_1, \ldots, \lambda_n\}$ such that $\sum_{i=1}^{n} \lambda_i x_i = x$. We claim that we may assume that each $\lambda_i > 0$. Indeed, since 0 is a weak internal point of Q, for each i there is $\alpha_i > 0$ such that $-\alpha_i x_i$ is in Q. Hence, if $\lambda_i < 0$, we may replace x_i by $-\alpha_i x_i$ and λ_i by $-\lambda_i / \alpha_i$ and leave the sum unchanged. Thus, we may assume $\lambda_i > 0$ for each i. Now set $\lambda = \sum_{i=1}^{n} \lambda_i$ and note that $\lambda^{-1} x$ is a convex combination of the x_i's, which therefore lies in Q. Using the fact that 0 is a weak internal point of Q, we get that $(0 - \frac{\beta}{\lambda} x) \in Q$ for some $\beta > 0$. Setting $\alpha = min\{\frac{1}{\lambda}, \frac{\beta}{\lambda}\}$, we have $\pm \alpha x \in Q$. Hence, Q is radial. $\square$

REMARK. If the hypotheses of Proposition 1.3 hold and, in addition, $\mathcal{X}$ is a linear topological space and Q is compact, then either $\mathcal{X}$ is finite dimensional or else it has uncountable linear dimension. Indeed, if there are finite-dimensional subspaces $\{\mathcal{Y}_n\}$ in $\mathcal{X}$ such that

$$\mathcal{X} = \cup \{ \mathcal{Y}_n : n = 1, 2, \ldots \},$$

then, by the Baire Category theorem [21,0.3.16], there is an n such that $\mathcal{Y}_n \cap Q$ contains a set U that is relatively open in Q. Hence, if we fix $u \in U$ and $x \in Q$, there is $\lambda > 0$ such that $(1 - \lambda)u + \lambda x \in U$. Thus,

$$x \in \lambda^{-1}(U - (1 - \lambda)u) \subset \mathcal{Y}_n.$$

Hence, $Q \subset \mathcal{Y}_n$ and $\mathcal{X} = \mathbb{R}$–span$(Q) \subset \mathcal{Y}_n$.

The following proposition will be used in the next section. Before stating it, we introduce another bit of notation. If v is a composite point of Q and $\mathcal{Y}$ is a subspace of $\mathcal{X}$, then we say that $\mathcal{Y}$ is **lineal** with respect to v if, for every element y in $\mathcal{Y}$, there is $\alpha > 0$ such that

$$\text{seg}(v + \alpha y, v - \alpha y) \subset Q.$$

(Although this notion depends on Q as well as v, this additional dependence is left implicit to avoid cumbersome notation).

PROPOSITON 1.4. *If v is a composite point in Q and we set*

$$\mathcal{Y} = \mathbb{R}\text{–span}(G(Q, v) - v),$$

then $\mathcal{Y}$ is lineal with respect to v. Also, if there is a positive integer n such that for each composite point v in Q we can find an n-dimensional real subspace $\mathcal{Y}$ that is lineal with respect to v, then $\dim_{\mathcal{F}}(Q) \geq n$.

PROOF. Fix a composite point v in Q. To simplify the notation write $C = G(Q, v) - v$. By part (2) of Theorem 1.2, v is a weak internal point of $G(Q, v)$. Hence, by part (2) of Proposition 1.1, 0 is a weak internal point of C, and so, by Proposition 1.3, C is radial in $\mathcal{Y}$. Now fix $y \in \mathcal{Y}$ and $\alpha > 0$ such that $\pm \alpha y \in C$. Since C is convex and contains 0, we have $\pm ty \in C$ for all $0 \leq t \leq \alpha$. Thus, $\text{seg}(v + \alpha y, v - \alpha y) \subset G(Q, v) \subset Q$ and so $\mathcal{Y}$ is lineal with respect to v.

For the latter assertion, fix a composite point v in Q, and suppose that there is an n-dimensional subspace $\mathcal{Y}$ that is lineal with respect to v. We claim that $\mathcal{Y}$ is contained in the linear span of $G(Q, v) - v$. To see this, select an element y in $\mathcal{Y}$, and use the fact that $\mathcal{Y}$ is lineal with respect to v to find $\alpha > 0$ such that $\text{seg}(v + \alpha y, v - \alpha y) \subset Q$. Since v is clearly an internal point of this line segment, we have $\text{seg}(v + \alpha y, v - \alpha y) \subset G(Q, v)$ by part (3) of Proposition 1.1. Hence, $\mathcal{Y}$ is contained in the linear span of $G(Q, v) - v$. Therefore, the dimension of the linear span of $G(Q, v) - v$ has dimension at least n, and so $\dim_{\mathcal{F}}(Q) \geq n$. $\square$

Suppose Ψ is an affine map of Q into a linear space and a Lyapunov theorem of type 1 holds for Ψ (i.e., $\Psi(Q) = \Psi(E(Q))$). In this case, if we fix $v \in Q$ and write

$$D_v = \Psi^{-1}(\Psi(v)),$$

then D_v is clearly convex and contains an extreme point $e \in E(Q)$. Thus, in order for a Lyapunov theorem of type 1 to hold, it is necessary and sufficient that D_v contain an extreme point of Q for every v in Q. If we assume that each D_v always contains an extreme point of itself (which will be true if Q is a compact subset of a locally convex space and Ψ is continuous), then we need only show that an extreme point of D_v is also extreme in Q. We call our next theorem (1.6) an abstract Lyapunov theorem because it provides a condition which ensures that this conclusion holds. We begin with some facts about extending affine maps. We are indebted to S. Simons for showing us the proof of the first assertion in the following proposition.

PROPOSITION 1.5. *If Q is a convex subset of a linear space $\mathcal{X}$, $0 \in Q$ and Φ is an affine map of Q into a linear space $\mathcal{Y}$ such that $\Phi(0) = 0$, then Φ extends uniquely to a linear map on $\mathbb{R}$-span$\{Q\}$.*

If, in addition, $\mathcal{X}$ and $\mathcal{Y}$ are locally convex spaces, Q contains an interior point of $\mathcal{X}$, and Φ is continuous on Q, then the extension is continuous.

PROOF. For the first assertion, let us begin by noting that, if $v \in Q$ and $0 < t < 1$, then, since $0 \in Q$ we have $tv \in Q$; thus, $\Phi(tv) = t\Phi(v)$ since Φ is affine. Let $\mathcal{Z}$ denote the real linear span of Q. We shall show that every $z \in \mathcal{Z}$ has the form $z = \alpha v - \beta w$, where $v, w \in Q, \alpha, \beta \geq 0$. A general element in $\mathcal{Z}$ has the form

$$z = \sum_{i=1}^{n} \alpha_i v_i - \sum_{j=1}^{m} \beta_j w_j,$$

where the v_i's and w_j's are in Q and the scalars α_i and β_j are non-negative. If we write $\alpha = \sum_{i=1}^n \alpha_i$ and $\beta = \sum_{i=1}^n \beta_i$, then (assuming α and β are positive) we have that

$$v = \frac{1}{\alpha} \sum_{i=1}^n \alpha_i v_i \quad \text{and} \quad w = \frac{1}{\beta} \sum_{j=1}^m \beta_j w_j$$

are each convex combinations of elements in Q, and so v and w are in Q. Hence, (even if α and/or $\beta = 0$) we get non–negative scalars α and β and elements v and w in Q such that

$$z = \alpha v - \beta w.$$

We now define our extension of Φ on z by

$$\Phi(z) = \alpha \Phi(v) - \beta \Phi(w).$$

We must show that this definition assigns a unique value to z. To see this, first note that, if $\pm v \in Q$, then we have

$$0 = \Phi(0) = \Phi(\tfrac{1}{2}v + \tfrac{1}{2}(-v)) = \tfrac{1}{2}\Phi(v) + \tfrac{1}{2}\Phi(-v),$$

and so

$$\Phi(-v) = -\Phi(v).$$

It follows readily that our definition assigns $\Phi(0) = 0$, no matter how 0 is represented. Now suppose that we have non–negative scalars $\alpha, \beta, \gamma, \delta$ and elements v, w, x, y in Q such that

$$\alpha v - \beta w = \gamma x - \delta y,$$

or, equivalently

$$(*) \qquad \alpha v + \delta y = \gamma x + \beta w.$$

Relabeling if necessary, we may assume $0 \leq \gamma + \beta \leq \alpha + \delta$. If $\alpha + \delta = 0$, then $\alpha = \delta = \gamma = \beta = 0$, so that $z = 0$, so $\Phi(z)$ is well-defined as above. Next, if $\gamma + \beta = 0$, then $\gamma = \beta = 0$, and we get

$$v = -\frac{\delta}{\alpha} y.$$

Relabeling again, we may assume $\frac{\delta}{\alpha} \leq 1$ and so, as noted above, $(\frac{\delta}{\alpha})y \in Q$. In this case, we have $\pm((\delta/\alpha)y) \in Q$ and therefore $\Phi(v) = -\frac{\delta}{\alpha}\Phi(y)$. Hence $\alpha \Phi(v) = -\delta \Phi(y)$ and our extension is well-defined in this case.

We are left with the case where $0 < \gamma + \beta \leq \alpha + \delta$. For this, we may now rewrite $(*)$ above as

$$(**) \qquad \frac{\alpha}{\alpha+\delta}v + \frac{\delta}{\alpha+\delta}y = \frac{\gamma+\beta}{\alpha+\delta}\left(\frac{\gamma}{\gamma+\beta}x + \frac{\beta}{\gamma+\beta}w\right).$$

Since $\frac{\gamma+\beta}{\alpha+\delta} \leq 1$, Φ is affine, $0 \in Q$ and both sides of $(**)$ lie in Q, we get

$$\frac{\alpha}{\alpha+\delta}\Phi(v) + \frac{\delta}{\alpha+\delta}\Phi(y) = \frac{\gamma+\beta}{\alpha+\delta}\left(\frac{\gamma}{\gamma+\beta}\Phi(x) + \frac{\beta}{\gamma+\beta}\Phi(w)\right),$$

and so $\alpha\Phi(v) - \beta\Phi(w) = \gamma\Phi(x) - \delta\Phi(y)$. Thus our extension is well-defined.

To see that Φ is linear on $\mathcal{Z}$, fix $z = \alpha v - \beta w$ in $\mathcal{Z}$ and a scalar γ. We have

$$\Phi(\gamma z) = \gamma\alpha\Phi(v) - \gamma\beta\Phi(w) = \gamma\Phi(z)$$

if γ is non-negative, and

$$\Phi(\gamma z) = -\gamma\beta\Phi(w) - (-\gamma\alpha)\Phi(v) = \gamma\Phi(z)$$

if γ is negative. Hence, Φ is homogeneous. For additivity, first consider a sum of the form $\alpha v + \alpha' v'$, where v and v' lie in Q and α and α' are non-negative scalars. There is a large positve number M such that $\frac{\alpha}{M}v$ and $\frac{\alpha'}{M}v'$ each lie in Q. Hence,

$$\frac{1}{2M}\Phi(\alpha v + \alpha' v') = \Phi(\tfrac{\alpha}{2M}v + \tfrac{\alpha'}{2M}v') = \tfrac{\alpha}{2M}\Phi(v) + \tfrac{\alpha'}{2M}\Phi(v'),$$

and so $\Phi(\alpha v + \alpha' v') = \alpha\Phi(v) + \alpha'\Phi(v')$. Thus, if $z = \alpha v - \beta w$ and $z' = \alpha' v' - \beta' w'$ are in $\mathcal{Z}$, then we have

$$\Phi(z + z') = \Phi(\alpha v + \alpha' v') - \Phi(\beta w + \beta' w') = \Phi(z) + \Phi(z'),$$

and so our extension is linear. It is clear from the definition of Φ on $\mathcal{Y}$ that this extension is unique.

Now suppose that $\mathcal{X}$ and $\mathcal{Z}$ are locally convex spaces, Q contains an interior point v, and Φ is continuous on Q. Recall that the topology on $\mathcal{X}$ is determined by a separating family of seminorms $\{p_k : k \in \kappa\}$ as follows. For each fixed $k \in \kappa$ and positive integer n write

$$V(p_k, n) = \{x \in \mathcal{X} : p_k(x) < \tfrac{1}{n}\}.$$

Each neighborhood of 0 contains a neighborhood of the form

$$(***) \qquad\qquad V(p_{k_1}, n_1) \cap \cdots \cap V(p_{k_m}, n_m);$$

i.e., the collection $\{V(p_k, n) : k \in \kappa, n \in \mathbb{N}\}$ forms a local subbase for the topology of $\mathcal{X}$ [39,1.34-1.38].

Fix an element $y \in \mathcal{Y}$ and a neighborhood U of $\Phi(y)$ in $\mathcal{Z}$. We must find a neighborhood of y that is mapped into U by Φ. As v is an interior point of Q, there is a neighborhood V of the form $(***)$ such that $v + V$ lies in Q. Replacing V by a subneighborhood, we may assume that $v + V + V \subset Q$. Also, there is $\alpha > 0$ such that $\alpha y \in V$ and therefore $v + \alpha y \in v + V \subset Q$. Since Φ is continuous on Q, and $\Phi(v + \alpha y) \subset \Phi(v) + \alpha U$, there is another neighborhood V' of the form $(***)$ such that

$$\Phi((v + \alpha y + V') \cap Q) \subset \Phi(v) + \alpha U.$$

Thus, if we write $V'' = V \cap V'$, then we have $v + \alpha y + V'' \subset v + V + V \subset Q$, and so

$$\Phi(v + \alpha y + V'') \subset \Phi(v) + \alpha U.$$

Hence,

$$\Phi(y + \tfrac{1}{\alpha}V'') \subset U,$$

as desired. $\square$

THEOREM 1.6 (ABSTRACT LYAPUNOV THEOREM). *If Q is a convex subset of a linear space $\mathfrak{X}$, $\dim_{\mathcal{F}}(Q) > n$, Ψ is an affine map of Q onto $\mathbb{C}^n$ and $v \in Q$, then every extreme point of D_v $(= \Psi^{-1}(\Psi(v)))$ is an extreme point of Q.*

PROOF. Let e denote an extreme point of D_v and suppose e is not an extreme point of Q. Write

$$R = G(Q, e) - e \qquad \text{and} \qquad \mathcal{Y} = \mathbb{R}\text{--span}(R).$$

Note that by Proposition 1.1, 0 is a weak internal point of R; thus, R is radial in $\mathcal{Y}$ by Proposition 1.3. Now define Φ on R by $\Phi(w - e) = \Psi(w) - \Psi(e)$. We have that Φ is affine on R and $\Phi(0) = \Phi(e - e) = 0$. Hence Φ extends to a linear map on $\mathcal{Y}$ by Proposition 1.5. Since $\dim_{\mathcal{F}}(Q) > n$ and e is a composite point of Q, the dimension of $\mathcal{Y}$ is greater than n, and so there is a nonzero element $y \in \mathcal{Y}$ such that $\Phi(y) = 0$. As R is radial in $\mathcal{Y}$, there is $\alpha > 0$ such that $\pm \alpha y \in R$. Thus we have

$$\alpha y = w - e \qquad \text{and} \qquad -\alpha y = w' - e$$

for some elements w and w' in $G(Q, e)$. Therefore

$$w - e = e - w' \qquad \text{and so} \qquad e = \tfrac{1}{2}(w + w').$$

On the other hand, we have $\Phi(\alpha y) = 0$, and so $\Psi(w) = \Psi(e)$; similarly $\Psi(w') = \Psi(e)$. Hence, w and w' lie in D_v and, since e is an extreme point of D_v, we get that $e = w = w'$. But then we have $y = 0$, which was explicitly ruled out. This is a contradiction and so e must be an extreme point in Q. $\square$

COROLLARY 1.7. *If Q is a compact subset of a locally convex space $\mathfrak{X}$ such that $\dim_{\mathcal{F}}(Q) > n$ and Ψ is an continuous affine map of Q into $\mathbb{C}^n$, then $\Psi(E(Q)) = \Psi(Q)$.*

PROOF. If $v \in Q$ and D_v is defined as in Theorem 1.6, then, since Ψ is continuous, D_v is compact. As D_v is also convex, it contains an extreme point e by the Krein-Milman theorem [39,p.70]. Since $\dim_{\mathcal{F}}(Q) > n$, we may apply Theorem 1.6 to conclude that $e \in E(Q)$. Since $e \in D_v$, we have $\Psi(e) = \Psi(v)$ and so $\Psi(E(Q)) = \Psi(Q)$. $\square$

Corollary 1.7 will be used in the next section to derive a type 1 Lyapunov theorem for nonatomic von Neumann algebras. (In the following paragraphs, Lemma 2.3 will be referred to; the reader might wish to read this result at this time). The next Corollary gives another application. In [47,p.956], the following version of the Phelps-Dye theorem is presented.

If (X, Σ, μ) is a nonatomic finite measure space and $\mathcal{E}$ is a finite dimensional subspace of $L^1(\mu)$ (viewed as a real vector space), then there is a subset $E \in \Sigma$ such that

$$\int_E f \, d\mu = \int_{X \setminus E} f \, d\mu$$

for each $f \in \mathcal{E}$.

The following gives a generalization of this theorem to a more abstract setting.

COROLLARY 1.8. *If $\mathfrak{X}$ is a normed linear space, $\mathcal{E}$ is a finite-dimensional subspace of $\mathfrak{X}$ and*

$$\dim_{\mathcal{F}}((\mathfrak{X}^*)_1) > dim(\mathcal{E}),$$

then the annihilator of $\mathcal{E}$ in $\mathfrak{X}^*$ contains an extreme point of $(\mathfrak{X}^*)_1$.

PROOF. Pick a basis $\{x_1, \ldots, x_n\}$ for $\mathcal{E}$ and define a mapping Ψ on $\mathfrak{X}^*$ by

$$\Psi(f) = \{f(x_1), \ldots, f(x_n)\}.$$

If we write $Q = (\mathfrak{X}^*)_1$, then Q is convex and weak* compact by the Alaoglu theorem [39,3.15, p.66] . As Ψ is clearly weak* continuous and $\dim_{\mathcal{F}}(Q) > n$, we may apply Corollary 1.6 and get that $\Psi(E(Q)) = \Psi(Q)$. Since $\dim_{\mathcal{F}}(Q) > n$, the dimension of $\mathfrak{X}^*$ is greater than n and so the kernel of Ψ is not empty. Hence there is an extreme point $f \in Q$ such that $\Psi(f) = 0$. Since $\{x_1, \ldots, x_n\}$ is a basis for $\mathcal{E}$, f is in the annihilator of $\mathcal{E}$. $\square$

In particular, if $\mathcal{M}$ is an abelian, nonatomic von Neumann algebra, then its predual, $\mathcal{M}_*$ is isomorphic to $L^1(\mu)$, where μ is nonatomic. If we write $\mathfrak{X}^*$ for the real linear span of the self-adjoint elements in $\mathcal{M}$, then, as we shall show in the next section (Lemma 2.3), $\dim_{\mathcal{F}}((\mathfrak{X}^*)_1) = \infty$. Thus, if $\mathcal{E}$ is a finite dimensional subspace of $\mathfrak{X}$, then Corollary 1.8 applies and we may find an extreme point s in $(\mathfrak{X})_1$ that annihilates the elements of $\mathcal{E}$. Such an extreme point must be a symmetry; i.e., $s = p - (1 - p)$ for some projection p in $\mathfrak{X}^*$. Translating this into measure–theoretic terms gives the version of the Phelps-Dye theorem mentioned above.

For another application we note that Theorem 1 of [24] follows quickly from Corollary 1.7 above since it is easy to show that the flat spots (i.e., the faces of non-zero dimension) on the unit ball of a Birnbaum-Orlicz space for a nonatomic measure must be infinite dimensional. The reader is referred to [24] for all the details of the (rather complicated) definitions of these spaces.

Thus far, our results have relied heavily on the fact that the range of Ψ has finite dimension. As the next result shows, however, it is possible to get an abstract Lyapunov theorem of type 2 for certain maps with infinite-dimensional range.

THEOREM 1.9. *If $\mathfrak{X}$ and $\mathcal{Y}$ are normed linear spaces, $\Psi : \mathfrak{X} \to \mathcal{Y}$ is a norm compact map and Q^* is a weak* compact, convex subset of $\mathcal{Y}^*$ with $\dim_{\mathcal{F}}(Q^*) = \infty$, then $\Psi^*(E(Q^*))$ is norm dense in $\Psi^*(Q^*)$.*

PROOF. Since Ψ is norm compact, for each fixed $\epsilon > 0$ there exist $x_1, \ldots, x_n$ in the unit ball of $\mathfrak{X}$ such that $\{\Psi(x_1), \ldots, \Psi(x_n)\}$ is ϵ-dense in $\Psi((\mathfrak{X})_1)$. If we now define $\Phi : \mathcal{Y}^* \to \mathbb{C}^n$ by

$$\Phi(y^*) = \{y^*(\Psi(x_1)), \ldots, y^*(\Psi(x_n))\},$$

then Φ is weak* continuous on $\mathcal{Y}^*$, and, since $\dim_{\mathcal{F}}(Q^*)$ is infinite, the hypotheses of Corollary 1.6 are satisfied. Thus, for each fixed $y^* \in Q^*$ there is an extreme point $e^* \in E(Q^*)$ such that $\Phi(e^* - y^*) = 0$. This means

$$e^*(\Psi(x_j)) - y^*(\Psi(x_j)) = \Psi^*(e^* - y^*)(x_j) = 0 \qquad \text{for } j = 1, \ldots, n.$$

Hence,

$$
\begin{aligned}
\|\Psi^*(e^*) - \Psi^*(y^*)\| &= \sup\{|\Psi^*(e^* - y^*)(x)| : \|x\| \le 1\} \\
&= \sup\{|(e^* - y^*)(\Psi(x))| : \|x\| \le 1\} \\
&\le \epsilon + \sup\{|(e^* - y^*)(\Psi(x_j))| : j = 1, \ldots, n\} \\
&= \epsilon.
\end{aligned}
$$

$\square$

Next, we present an example that shows that *weak* compactness of Ψ is not sufficient in Theorem 1.9.

EXAMPLE 1.10. Set $\mathcal{X} = \ell^1(\mathbb{N})$ and $\mathcal{Y} = L^1([0,1], \lambda)$, where λ is Lebesgue measure (so that $\mathcal{Y}^* = L^\infty([0,1])$). Also, fix a norm dense sequence $\{f_n\}$ in

$$
S = \{f \in \mathcal{Y} : 0 \le f \le 1\}
$$

and define $\Psi : \mathcal{X} \to \mathcal{Y}$ by

$$
\Psi(\langle \alpha_n \rangle) = \sum_{n=1}^{\infty} \alpha_n f_n.
$$

Recall that a map Ψ from $\mathcal{X}$ to $\mathcal{Y}$ is **weakly compact** if the image of the unit ball of $\mathcal{X}$ under Ψ is a weakly precompact set in $\mathcal{Y}$ [21,p.618]. By [1,II.3] S is a weakly compact set in L^1. It follows that Ψ is a weakly compact map. Now write $Q^* = \{f \in \mathcal{Y}^* : 0 \le f \le 1\}$ and suppose that there exists p in $E(Q^*)$ such that

$$
(*) \qquad \left\| \Psi^*\left(p - \frac{1}{2}1\right) \right\|_\infty < \frac{1}{5}.
$$

Since $p \in E(Q^*)$, p is the characteristic function of some measurable subset of $[0,1]$; so $p^2 = p$. It is straight-forward to check that the nth coordinate of $\Psi^*(f)$ is $\int f f_n \, d\lambda$. Hence, by $(*)$, we get that

$$
(**) \qquad \left| \int \left(p f_n - \frac{1}{2} f_n\right) d\lambda \right| \le \frac{1}{5} \qquad \text{for } j = 1, 2, \ldots .
$$

Since $\{f_n\}$ is norm dense in S it follows that $(**)$ holds for arbitrary $f \in S$. Successively taking $f = p$ and $f = 1 - p$ we get

$$
\left| \int \left(\frac{1}{2} p\right) d\lambda \right| \le \frac{1}{5} \qquad \text{and} \qquad \left| \int \left(-\frac{1}{2}(1 - p)\right) d\lambda \right| \le \frac{1}{5}.
$$

Hence,

$$
\frac{1}{2} = \left| \int \frac{1}{2} 1 \, d\lambda \right| = \left| \int \left(\frac{1}{2} p + \frac{1}{2}(1 - p)\right) d\lambda \right| \le \frac{2}{5},
$$

which is impossible. Thus , we have that

$$
\left\| \Psi^*\left(p - \frac{1}{2}1\right) \right\| \ge \frac{1}{5}
$$

for every extreme point p and so no Lyapunov theorem of type 2 can hold for this Ψ^*, despite the fact that (by Lemma 2.3) all the hypotheses of Theorem 1.9 hold except that weak compactness is substituted for norm compactness.

Lastly, we show that the hypotheses of Theorem 1.9 do not support a type 1 conclusion.

Example 1.11. If we redefine the map Ψ in the previous example by

$$\Psi(\langle \alpha_n \rangle) = \sum_{n=1}^{\infty} 2^{-n} \alpha_n f_n \,,$$

then it is easy to check that Ψ is a compact map. Nevertheless, we may use an argument like that in the last example to show that $\Psi^*(E(Q^*)) \neq \Psi^*(Q)$.

§2. LYAPUNOV THEOREMS FOR
NONATOMIC VON NEUMANN ALGEBRAS

Throughout this section we shall be concerned with the case where $\mathcal{M}$ is a nonatomic von Neumann algebra. The main result of this section, Theorem 2.5, states that a Lyapunov theorem of type 1 holds for weak* continuous, affine maps from faces in $(\mathcal{M})_1$ to $\mathbb{C}^n$. The facial structure of the unit ball of an arbitrary von Neumann algebra has recently been studied [4], [20]. Let us begin by reviewing the salient results. We shall abuse notation slightly by calling faces of $(\mathcal{M}_+)_1$ **positive faces** of $(\mathcal{M})_1$ and faces in $(\mathcal{M}_{sa})_1$ **self-adjoint faces** of $(\mathcal{M})_1$.

THEOREM 2.1. *If $\mathcal{M}$ is a von Neumann algebra then the following assertions are true:*

(1) *If F is a positive, weak* closed face in $(\mathcal{M})_1$, then F has the form*

$$F = [p, q] = \{\, a : p \le a \le q \,\}$$

for a unique pair of projections p and q in $\mathcal{M}$ with $p \le q$.

(2) *If F is a self-adjoint, weak* closed face in $(\mathcal{M})_1$, then F has the form*

$$F = [s, t] = \{\, x : s \le x \le t \,\}$$

for a unique pair of symmetries in $\mathcal{M}$ with $s \le t$.

(3) *If p and q are projections in $\mathcal{M}$ and F is a weak* closed face in $(p\mathcal{M}q)_1$, then F has the form*

$$F = v + (p - vv^*)(p\mathcal{M}q)_1(q - v^*v)$$

for a unique partial isometry v in $p\mathcal{M}q$.

PROOF. Statements (1) and (2) are just Theorems 2.2 and 3.1 of [4] (also see [20,Theorem 4.1]). Statement (3) follows from Theorem 6.6 and Remark 4.5 of [4]. □

As the next proposition shows, one may always reduce the study of weak* closed faces in $(\mathcal{M}_+)_1, (\mathcal{M}_{sa})_1$ or $(p\mathcal{M}q)_1$ to the study of the study of $(\mathcal{N}_+)_1$ or $(p'\mathcal{N}q')_1$ in an hereditary von Neumann subalgebra $\mathcal{N}$.

PROPOSITION 2.2. *If $\mathcal{M}$ is a von Neumann algebra, then the following state-ments are true:*

(1) *If F is a positive, weak* closed face in $(\mathcal{M})_1$, then there is an heredi-tary von Neumann subalgebra $\mathcal{N}$ of $\mathcal{M}$ and a weak* continuous affine isomorphism of F onto $(\mathcal{N}_+)_1$.*

(2) *If F is a self-adjoint, weak* closed face in $(\mathcal{M})_1$, then there is an hered-itary von Neumann subalgebra $\mathcal{N}$ of $\mathcal{M}$ and a weak* continuous affine isomorphism of F onto $(\mathcal{N}_+)_1$.*

(3) *If p and q are projections in $\mathcal{M}$ and F is a weak* closed face in $(p\mathcal{M}q)_1$, then there projections p' and q' in $\mathcal{M}$ and a weak* continuous affine isomorphism of F onto $(p'\mathcal{M}q')_1$.*

PROOF. If $F = [p, q]$ is a positive face, then write $r = q - p$, $\mathcal{N} = r\mathcal{M}r$ and define a map Φ on F by $\Phi(a) = a - p$. It is clear that Φ has the desired properties and (1) holds. For (2), note that the map $x \mapsto (x - 1)/2$ is a weak* continuous, affine isomorphism of F onto a weak* closed face in $(\mathcal{M}_+)_1$. Hence, composing this map with the map defined in (1) we get the desired conclusion. Finally, suppose F is a face in $(p\mathcal{M}q)_1$. By part (3) of Theorem 2.1, we have

$$F = v + ((p - vv^*)(p\mathcal{M}q)_1(q - v^*v)$$

for a unique partial isometry $v \in \mathcal{M}$. If we write $p' = p - vv^*$ and $q' = q - v^*v$, then the map defined by $x \mapsto x - v$ has the required properties. $\square$

Since we shall rely on Corollary 1.7 to obtain our main theorem in this section, the utility of the following two lemmas is clear.

LEMMA 2.3. *If $\mathcal{M}$ is a nonatomic von Neumann algebra and F is a positive (resp. self-adjoint) weak* closed face in $(\mathcal{M})_1$, then $\dim_{\mathcal{F}}(F) = \infty$.*

PROOF. By part (2) of Proposition 2.2 we need only prove the assertion for positive faces. Moreover, by part (1) of Proposition 2.2, we may assume $F = (\mathcal{N}_+)_1$. Since $\mathcal{N}$ is an hereditary von Neumann subalgebra of $\mathcal{M}$, $\mathcal{N}$ is also nonatomic. Thus, to establish the Lemma, we may reduce to the case where $F = (\mathcal{M}_+)_1$. So, fix a composite point $a \in (\mathcal{M}_+)_1$ and note that, by Proposition 1.4, we need only find an infinite-dimensional real subspace of $\mathcal{M}$ that is lineal with respect to a. Since a is a composite point, there is $0 < \delta < 1/2$ such that the spectral projection of a corresponding to the interval $(\delta, 1 - \delta)$ is nonzero. Denote this projection by p and write $\mathcal{Y} = (p\mathcal{M}p)_{sa}$. Since $\mathcal{M}$ is nonatomic and $p \neq 0$, $\mathcal{Y}$ is infinite-dimensional. To see that $\mathcal{Y}$ is lineal with respect to a, fix $y \neq 0$ in $\mathcal{Y}$ and put $\gamma = \delta/\|y\|$. If $0 < t \leq \gamma$, then ty is self-adjoint and

$$\|a + ty\| = \max\{\|(1 - p)a\|, \|ap + ty\|\} \leq \max\{\|(1 - p)a\|, \|ap\| + \delta\} \leq 1.$$

Hence, $a + ty \leq 1$. Similarly, $1 - (a + ty) \leq 1$ and so $a + ty \in (\mathcal{M}_+)_1$. Replacing y by $-y$ and repeating the argument, we get $0 \leq a - ty \leq 1$; so, $a \pm ty \in (\mathcal{M}_+)_1$ and $\mathcal{Y}$ is lineal with respect to a. $\square$

LEMMA 2.4. *If $\mathcal{M}$ is a nonatomic von Neumann algebra, p and q are nonzero projections in $\mathcal{M}$ and F is a weak* closed face in $(p\mathcal{M}q)_1$, then F has infinite facial dimension.*

PROOF. By part (3) of Proposition 2.2, it suffices to prove the assertion in the case where $F = (p\mathcal{M}q)_1$. So, fix a nonzero composite point x in $(p\mathcal{M}q)_1$ and observe that, as in the proof of Lemma 2.3, it is enough to produce an infinite-dimensional real subspace that is lineal with respect to x. Let $v|x|$ denote the polar decomposition of x and observe that $|x|q = q|x| = |x|$. Since we also have $px = x$, we get $x = pvq|x|$, and so we may assume $v = pvq$. We now consider 2 cases.

First suppose $|x|$ is not a projection. In this case, if we write

$$r = v^*v \qquad \text{and} \qquad \mathcal{Z} = \mathbb{R}\text{--span}(G((r\mathcal{M}r)_1, |x|) - |x|),$$

then $\mathcal{Z}$ is lineal with respect to $|x|$ by Proposition 1.4 and, since $r \neq 0$, $\mathcal{Z}$ has infinite dimension as in the proof of Lemma 2.3. If $z \in \mathcal{Z}$, then $z^*v^*vz = z^*rz = z^*z$. Thus, if we write $\mathcal{Y} = v\mathcal{Z}$, then $\mathcal{Y}$ is also infinite-dimensional. Now fix $y = vz \in \mathcal{Y}$. As $\mathcal{Z}$ is lineal with respect to $|x|$, there is $\delta > 0$ such that $\text{seg}(|x| + \delta z, |x| - \delta z) \subset \mathcal{Z}$. It follows that $\text{seg}(x + \delta y, x - \delta y) \subset \mathcal{Y}$, and so $\mathcal{Y}$ is lineal with respect to x.

Finally, suppose $|x|$ is a projection. In this case, x is a partial ismometry. Since x is a composite point, if we write $\mathcal{Y} = (p - xx^*)\mathcal{M}(q - x^*x)$, then $\mathcal{Y} \neq 0$ [40,1.6.5] and, since $\mathcal{M}$ is nonatomic, $\mathcal{Y}$ has infinite dimension. Finally, if $y \in \mathcal{Y}$ and $\|y\| \leq 1$, then $\|x \pm y\| \leq 1$ and therefore $\mathcal{Y}$ is lineal with respect to x. □

We are now ready to present the main theorem of this section. We note that special cases of part 1 have been established previously by Azarnia and Wright [9,Cor. 3] and Choda, et al. [15,Theorem 1].

THEOREM 2.5 (LYAPUNOV THEOREM FOR NONATOMIC VON NEUMANN ALGEBRAS). *If F is a weak* closed subset of a nonatomic von Neumann algebra $\mathcal{M}$ and Ψ is a weak* continuous affine map of F into $\mathbb{C}^n$, then the following statements are true.*

(1) *If F is a positive face in $(\mathcal{M})_1$ and $a \in F$, then there is a projection $p \in F$ such that p commutes with a and $\Psi(p) = \Psi(a)$.*

(2) *If F is a self-adjoint face in $(\mathcal{M})_1$ and $x \in F$, then there is a symmetry $s \in F$ such that s commutes with x and $\Psi(s) = \Psi(x)$.*

(3) *If F is a face of $(\mathcal{M})_1$ and x is a normal element in F, then there is a unitary $u \in F$ such that u commutes with x and $\Psi(u) = \Psi(x)$.*

(4) *If p and q are projections in $\mathcal{M}$, F is a face of $(p\mathcal{M}q)_1$ and $x \in F$, then there is an element w in $E(F)$ such that $\Psi(w) = \Psi(x)$.*

PROOF. We prove parts (1), (2) and (3) simultaneously. Select a MASA $\mathcal{N}$ in $\mathcal{M}$ that contains a (or x for parts (2) or (3)). Since $\mathcal{M}$ is nonatomic, $\mathcal{N}$ is also nonatomic and so, in each case, $F \cap \mathcal{N}$ has infinite facial dimension by the previous Lemmas. Also, in each case, $F \cap \mathcal{N}$ is a weak* compact, convex subset of $\mathcal{N}$. Hence, Corollary 1.7 applies and there is an extreme point $e \in E(F \cap \mathcal{N})$ such that $\Psi(e) = \Psi(a)$ (or $\Psi(e) = \Psi(x)$ for parts (2) or (3)). Thus, in each case, e commutes with the corresponding element. Since e is an extreme point, we have that $e = p$ is a projection in (1), $e = s$ is a symmetry in (2) and $e = u$ is a unitary in (3), as desired. For (4), note that F is weak* compact and convex and $dim_{\mathcal{F}}(F) = \infty$ by Lemma 2.4. Hence, Corollary 1.7 also applies in this case to give the required extreme point. $\square$

Using Theorem 2.5 and the method introduced by Lindenstrauss, Lyapunov's original theorem is now easy to establish.

COROLLARY 2.6: (LYAPUNOV'S THEOREM). *If* $(X, \mathcal{S})$ *is a measurable space and* $\mu : \mathcal{S} \to \mathbb{C}^n$ *is a nonatomic vector-valued measure, then the range of* μ *is convex and compact.*

PROOF. Let $\mu_1, \ldots, \mu_n$ denote the coordinate measures determined by μ and write $|\mu| = |\mu_1| + \cdots + |\mu_n|$ We have that $|\mu|$ is a positive, finite, nonatomic measure. Write $\mathcal{M} = L^\infty(X, \mathcal{S}, |\mu|)$ and define Ψ on $\mathcal{M}$ by

$$\Psi(f) = <\textstyle\int f \, d\mu_1, \ldots, \int f \, d\mu_n> .$$

As each μ_j is absolutely continuous with respect to $|\mu|$, by the Radon-Nykodym theorem, $d\mu_j = f_j d|\mu|$ for some f_j in $L^1(X, \mathcal{S}, |\mu|)$. Hence, Ψ is weak* continuous and so $\Psi((\mathcal{M}_+)_1))$ is convex and compact. As $\mathcal{M}$ is nonatomic, part (1) of Theorem 2.5 applies. Thus, for each f in $(\mathcal{M}_+)_1$ there is a projection p such that $\Psi(p) = \Psi(f)$. Since each projection is given by the characteristic function of a measurable set, we get $\Psi((\mathcal{M}_+)_1) = \mu(\mathcal{S})$. Hence, $\mu(\mathcal{S})$ is convex and compact. $\square$

EXAMPLE 2.7. Lyapunov himself [34] showed that if we drop the assumption that the range of Ψ is finite-dimensional, then the conclusions of Theorem 2.5 are, in general, false even in the abelian case. In fact, as we now show, there are maps with infinite-dimensional range for which even a type 4 Lyapunov theorem fails to hold.

Let $\mathcal{M}$ denote an arbitrary σ-finite von Neumann algebra and fix a sequence $\{f_j\}$ that is dense in the unit ball of the predual of $\mathcal{M}$. If we define $\Psi : \mathcal{M} \to \ell^\infty(\mathbb{N})$ by $\Psi(x) = \{f_j(x)\}$, then Ψ is clearly weak* to weak* continuous. It is also clear that Ψ is isometric; hence, no Lyapunov theorem of any type can hold for Ψ.

§3. LYAPUNOV THEOREMS FOR C^*–ALGEBRAS

In this section we consider the case where Ψ is a norm continuous map of a C^*–algebra A into $\mathbb{C}^n$. Recall that we denote the set of projections (resp. unitaries) in A by P_A (resp. U_A) or just P (resp. U) if the ambiant algebra is clear. We begin by studying the weak* closure of $P_\mathcal{M}$ and $U_\mathcal{M}$ for a von Neumann algebra $\mathcal{M}$. If the finite type I part of $\mathcal{M}$, $\mathcal{M}_{fin}$, is $\{0\}$, then we show in Theorem 3.2 that P (resp. U) is weak* dense in $(\mathcal{M}_+)_1$ (resp. $(\mathcal{M})_1$). From this it is easy to show that, if A is unital, then $\Psi(U_A)$ is dense in $\Psi((A)_1)$. If A also has "sufficiently many" projections, then we have that $\Psi(P_A)$ is dense in $\Psi((A_+)_1)$. We begin with a technical lemma.

LEMMA 3.1. *If N is a weak* neighborhood of a partial isometry v in a von Neumann algebra $\mathcal{M}$, then there are partial isometries w and u in $\mathcal{M}$ such that $w \in N$ and $u + w$ is unitary.*

PROOF. We have the decomposition $\mathcal{M} = \mathcal{M}_1 \oplus \mathcal{M}_2$ where $\mathcal{M}_1$ is semifinite and $\mathcal{M}_2$ is purely infinite (i.e. of type III); clearly it is enough to prove the theorem for each summand separately.

First suppose $\mathcal{M}$ is semifinite. In this case there is a net $\{p_\alpha\}$ of finite projections that converges to v^*v in the weak* topology and such that $p_\alpha \leq v^*v$ for each α. Hence $vp_\alpha \in N$ for some fixed α. Using this α, write $p = p_\alpha$, $w = vp$ and $q = ww^*$. We have $w \in N$ by definition, and, since $p = w^*w$, p and q are equivalent finite projections. Hence, $1 - p$ and $1 - q$ are equivalent [27,6.9.7] and, if u denotes a partial isometry that implements this equivalence, then $u + w$ is unitary, as desired.

Next suppose that $\mathcal{M}$ is purely infinite. Observe that by replacing N by a smaller neighborhood if necessary, we may assume that it has the form

$$N = \{a \in \mathcal{M} : |f_i(a - v)| < \epsilon, \ i = 1, \ldots, n\},$$

where the f_i's are weak* continuous linear functionals. Since each f_i is a linear combination of states, we may replace N by an even smaller neighborhood if necessary and assume that each f_i is a state. Let $f = \sum f_i$ and note that $\|f\| = f(1) = n$.

Write $r = v^*v$ and fix a positive integer m such that $m > n^2 \epsilon^{-2}$. Since $\mathcal{M}$ is purely infinite, we may repeatedly use the halving lemma [27,6.3.3] to obtain orthogonal equivalent projections $r_1, \ldots, r_m$ such that

$$r_1 + \cdots + r_m = r \quad \text{and} \quad r_j \sim r \sim r - r_j \quad \text{for each} \ j.$$

Relabeling if necessary, we may assume that $f(r_m) \leq f(r_j)$ for each j and so

$$f(r_m) \leq \frac{f(r)}{m} < \frac{n}{m} < \frac{\epsilon^2}{n}.$$

Thus, using Cauchy–Schwarz, we have

$$|f(v(r - r_m) - v)| = |f(vr_m)| \leq \sqrt{f(vv^*)f(r_m)} < \epsilon,$$

and so $v(r - r_m)$ is in N. Write $p = r - r_m$ and observe that

$$1 - p \geq r - p = r_m \sim p.$$

Since $1 = p \vee (1 - p)$, we get $1 \sim 1 - p$ by [27,6.9.4]. Using a similar argument we may find a projection q such that $q \leq vpv^*$, $qvp \in N$, $q \sim vpv^* \sim p$ and $1 \sim 1 - q$. Hence, $1 - p$ and $1 - q$ are equivalent via a partial isometry u. If we put $w = qvp$, then u and w have the desired properties. $\square$

As noted in the introduction, we have the decomposition

$$\mathcal{M} = \mathcal{M}_{fin} \oplus \mathcal{M}_{inf} \oplus \mathcal{M}_{na}.$$

If T is a subset of $\mathcal{M}$, we write

$$T_{fin} = T \cap \mathcal{M}_{fin}, \quad T_{inf} = T \cap \mathcal{M}_{inf} \quad \text{and} \quad T_{na} = T \cap \mathcal{M}_{na}.$$

Also we let

$$(T)^{-wk^*}$$

denote the closure of T in the weak* topology. We note that special cases of part iii) of the following theorem were proved by Dye [18] and Halmos [25].

THEOREM 3.2. *If $\mathcal{M}$ is a von Neumann algebra, then*
  i) $(P_{fin})^{-wk^*} = P_{fin}$ *and* $(U_{fin})^{-wk^*} = U_{fin}$,
  ii) $(P_{inf} \oplus P_{na})^{-wk^*} = ((\mathcal{M}_{inf} \oplus \mathcal{M}_{na})_+)_1$ *and*
  iii) $(U_{inf} \oplus U_{na})^{-wk^*} = (\mathcal{M}_{inf} \oplus \mathcal{M}_{na})_1$.

PROOF. If z denotes a central projection such that $z\mathcal{M}$ is a finite type I factor, then Pz and Uz are weak* closed, so it is clear that i) holds.

To establish ii), note that if H is infinite–dimensional and $P(H)_{inf}$ denotes the projections of infinite rank in $B(H)$, then

$$(P(H)_{inf})^{-wk^*} = (B(H)_+)_1$$

by [27,5.7.13]. Since $\mathcal{M}_{inf}$ is the direct sum of factors, each isomorphic to $B(H)$ (where the cardinality of H may vary), we get that

$$(P_{inf})^{-wk^*} = ((\mathcal{M}_{inf})_+)_1.$$

Now fix a positive element a in $(\mathcal{M}_{na})_1$ and a weak* neighborhood N of $\mathcal{M}_{na}$ that contains a. Replacing N by a smaller neighborhood if necessary, we may assume N has the form

$$N = \{x \in \mathcal{M}_{na} : |f_i(x - a)| < \epsilon, \ i = 1, \dots, n\},$$

where the f_i's are weak* continuous linear functionals. Define Ψ on $\mathcal{M}_{na}$ by

$$\Psi(x) = (f_1(x), \dots, f_n(x)).$$

Since $\mathcal{M}_{na}$ is nonatomic and Ψ is weak* continuous, Theorem 2.5 applies. Hence, there is a projection p in $\mathcal{M}_{na}$ such that $\Psi(p) = \Psi(a)$, and therefore $p \in N$. Hence,

$$(P_{na})^{-wk^*} = ((\mathcal{M}_{na})_+)_1,$$

and so ii) holds.

To establish iii) first note that

$$(U_{inf})^{-wk^*} = (\mathcal{M}_{inf})_1$$

by [27,5.7.13]; so, we may again restrict our attention to $\mathcal{M}_{na}$. Fix x in $(\mathcal{M}_{na})_1$ and let N denote a weak* neighborhood that contains x. Defining Ψ as before and applying part iv) of Theorem 2.5 (with $p = q = 1$), we get that N contains an extreme point v of $(\mathcal{M}_{na})_1$. By [40,1.6.5], v is a partial isometry. Hence, by Lemma 3.1, there are partial isometries w and u such that $w \in N$ and $u + w$ is unitary. Write $p = w^*w$. By part ii) of this Theorem, there is a net $\{p_\alpha\}$ of projections that converges to $(1-p)/2$ in the weak* topology. Hence, if we write $u_\alpha = 1 - 2p_\alpha$, then each u_α is unitary and the net $\{u_\alpha\}$ converges to p. Since $up = 0$ and $wp = w$, we get that the net $\{(u + w)u_\alpha\}$ converges to w. Hence, for some α, the unitary $(u + w)u_\alpha$ lies in N. Thus, iii) is true and the proof is complete. $\square$

The last theorem suggests that the lack of finite dimensional direct summands in a von Neumann algebra allows approximations which would otherwise be impossible. Tsui and Wright [45,Lemma 2.8] have investigated such algebras. We now record the definition of the corresponding property for C^*-algebras.

DEFINITION 3.3. *A C^*-algebra is said to be* **weakly infinite** *if it does not admit any nonzero finite-dimensional representations.*

Note that A is weakly infinite if and only if its second dual A^{**} has no finite type I direct summand. This simple observation provides the basis for our next 2 results.

THEOREM 3.4. *If A is a unital weakly infinite C^*–algebra, then the set U of unitaries is weakly dense in $(A)_1$.*

PROOF. Write $\mathcal{M} = A^{**}$, fix a in $(A)_1$ and let W denote a weak neighborhood of a. Observe that W has the form $W = N \cap A$, where N is a weak* neighborhood of a, where we identify A with its cannonical embedding in its second dual $\mathcal{M}$

[42, p.122]. Since A has no representations of finite rank, we have $\mathcal{M}_{fin} = \{0\}$. So, by part iii) of Theorem 3.2 there is a unitary u in $\mathcal{M}$ that lies in N. By [37,2.7.5] the unitaries of A are strong* dense in the unitaries of $\mathcal{M}$. Hence, there is a unitary element v of A that lies in W. $\square$

Recall that a C^*–algebra A is said to have **property FS** if each self–adjoint element can be approximated arbitrarily well in norm by a finite linear combination of commuting projections [11] and that A has **real rank zero** if the invertible self–adjoint elements are dense in the set of all self–adjoint elements [14]. It is an easy exercise to show that these 2 notions are equivalent for unital C^*–algebras [14].

THEOREM 3.5. *If A is a weakly infinite C^*–algebra that has property FS, then the set P of projections in A is weakly dense $(A_+)_1$.*

PROOF. Write $\mathcal{M} = A^{**}$. By using part ii) of Theorem 3.2 and arguing as in the proof of Theorem 3.3, we get that the projections of $\mathcal{M}$ are weak* dense in $(\mathcal{M}_+)_1$. By [1,III.2], each projection in A^{**} can be strong* approximated by projections in A. It follows, as in the proof of Theorem 3.3, that P is weakly dense $(A_+)_1$. $\square$

Recall that S stands for the set of symmetries in an operator algebra. We now present the main result of this section. Note that although we obtain conclusions that are generally analogous to those of Theorem 2.5, we no longer can find a commuting approximant.

THEOREM 3.6. (LYAPUNOV THEOREM FOR C^*–ALGEBRAS). *Suppose that A is a weakly infinite C^*–algebra and that Ψ is a norm continuous linear map of A into $\mathbb{C}^n$.*

   i) *If A is unital, then $\Psi(U)$ is dense in $\Psi((A)_1)$.*
   ii) *If A has property FS, then $\Psi(P)$ is dense in $\Psi((A_+)_1)$.*
   iii) *If A is unital and has property FS, then $\Psi(S)$ is dense in $\Psi((A_{sa})_1)$.*

PROOF. For i) observe that since Ψ is norm continuous, it is weakly continuous. By theorem 3.4, U is weakly dense in $(A)_1$ and so its image $\Psi(U)$ is weakly dense in $\Psi((A)_1)$. Because $\mathbb{C}^n$ is finite dimensional, $\Psi(U)$ is norm dense in $\Psi((A)_1)$.

Part ii) follows from Theorem 3.5 and a similar argument . Finally, iii) is immediate from ii) because $(A_{sa})_1 = 2(A_+)_1 - 1$ and $S = 2P - 1$. $\square$

The next three examples show that, for C*-algebras, the best we can expect to get is a Lyapunov theorem of type 2. Moreover, in the finite-dimensional case, not even a type 2 result is possible.

EXAMPLE 3.7. If A is a separable C*-algebra, then its K–groups are countable [11]. If A admits a trace τ, then

$$\tau(P) \subset \tau_*(K_0(A)),$$

so $\tau(P)$ is countable. Thus, if A is a simple AF algebra with trace τ, then A is weakly infinite and has real rank zero, but $\tau(P) \neq \tau((A_+)_1)$.

EXAMPLE 3.8. Write $A = B(H)$ and let f denote a normal pure state on A. We have then that f is supported by a projection p of rank one. Let g be a normal state that is faithful on $1 - p$ and such that $g(1 - p) = 1$. If we define Ψ by $\Psi(a) = (f(a), g(a))$, then $\Psi(1 - \frac{1}{2}p) = (\frac{1}{2}, 1)$. If we fix x in $\Psi^{-1}((\frac{1}{2}, 1)) \cap (A)_1$, then we have $g(x) = 1$. Since g is faithful on $1 - p$, it follows that $(1 - p)x = x(1 - p) = 1 - p$. Using the fact that $f(1 - p) = 0$, we get that $x = \frac{1}{2}p + 1 - p = 1 - \frac{1}{2}p$. Thus, $\Psi(U) \neq \Psi((A)_1)$ and $\Psi(P) \neq \Psi((A_+)_1)$

EXAMPLE 3.9. Write

$$A = M_2(\mathbb{C}),$$

the 2 by 2 complex matrices, and take Ψ to be the map that sends a matrix to its diagonal. In this case $\Psi(U)$ is not dense in $\Psi((A)_1)$ and $\Psi(P)$ is not dense in $\Psi((A_+)_1)$. Indeed, suppose that we have a unitary

$$u = \begin{bmatrix} \alpha & \beta \\ \gamma & \delta \end{bmatrix}$$

and a projection

$$p = \begin{bmatrix} t & \epsilon \\ \epsilon^* & s \end{bmatrix}$$

in $M_2(\mathbb{C})$. Since u is unitary,

$$|\alpha|^2 + |\beta|^2 = |\delta|^2 + |\beta|^2 = 1,$$

and so $|\alpha| = |\delta|$. Hence, if we write

$$q = \begin{bmatrix} 1 & 0 \\ 0 & 0 \end{bmatrix},$$

then $q \in (M_2(\mathbb{C}))_1$ and

$$\|\Psi(u - q)\| \geq \frac{1}{2}$$

for every unitary u in $M_2(\mathbb{C})$. Also, for the projection p above, we have that $s + t$ is 0, 1 or 2, and it follows that, if

$$a = \begin{bmatrix} 1/2 & 0 \\ 0 & 0 \end{bmatrix},$$

then

$$\|\Psi(p - a)\| \geq \frac{1}{4}$$

for every projection p in $\mathcal{M}$.

A von Neumann algebra $\mathcal{M}$ is weakly infinite if and only if $\mathcal{M}$ does not have a finite type I direct summand. Indeed, it is clear that if $\mathcal{M}$ has a finite type I direct summand, then $\mathcal{M}$ is not weakly infinite. Conversely, if $\mathcal{M}$ does not have such a summand then it decomposes as the direct sum of a properly infinite algebra and an algebra of type II$_1$. Using the halving lemmas [27,6.3.3] and [27,6.5.6], we may write the identity as the sum of n orthogonal equivalent projections for any positive integer n. It follows that $\mathcal{M}$ does not admit any representations of finite rank; i.e., $\mathcal{M}$ is weakly infinite. Since it is trivial that a von Neumann algebra is unital and has real rank zero, Theorem 3.5 immediately yields a Lyapunov theorem for weakly infinite von Neumann algebras. In fact, in this case, we also can get a conclusion similar to that of part 4 of Theorem 2.5. Since the argument for this will be needed in several other results, we isolate it here in the next 2 lemmas.

Recall that for each each element x in a von Neumann algebra there is a unique partial isometry v such that v^*v is the projection onto the closure of the range of $|x|$ and $x = v|x|$ [37,2.2.9]. We call this v the **minimal partial isometry** for the polar decomposition of x.

LEMMA 3.10. *If x is an element in a von Neumann algebra $\mathcal{M}$, then the following statements are true.*

(1) $F = \{w \in (\mathcal{M})_1 : x = w|x|\}$ *is a weak* closed face in* $(\mathcal{M})_1$.
(2) *If v is the minimal partial isometry for the polar decomposition of x, then $F = v + (1 - vv^*)(\mathcal{M})_1(1 - v^*v)$.*

PROOF. As noted in Remark 4.5 of [4], both assertions will follow if we can show that $w \in F$ if and only if $wv^* = vv^*$. If the latter condition holds, then

$$w|x| = w(v^*v)|x| = (wv^*)v|x| = vv^*v|x| = v|x| = x$$

and so $w \in F$.

For the converse, suppose $w \in F$. We may asssume that $\mathcal{M}$ acts as operators on a Hilbert space H. Fix $\xi \in H$ and write $\eta = |x|\xi$. We have

$$w\eta = w|x|\xi = v|x|\xi = v\eta.$$

Hence w and v agree on the closure of the range of $|x|$ and so $wv^*v = vv^*v$. Thus, $wv^* = wv^*vv^* = vv^*vv^* = vv^*$, as desired. $\square$

LEMMA 3.11. *If p and q are projections in a von Neumann algebra $\mathcal{M}$, then the following statements are true.*

(1) *If $x \in p\mathcal{M}q$, then there is an extreme point w of $(p\mathcal{M}q)_1$ such that $x = w|x|$.*
(2) *If w is an extreme point of $(p\mathcal{M}q)_1$ and u is a unitary in $(q\mathcal{M}q)_1$, then wu is an extreme point of $(p\mathcal{M}q)_1$.*

PROOF. Fix $x \in (p\mathcal{M}q)_1$ and let F and v be as in Lemma 3.10. We have that $F_{pq} = F \cap (p\mathcal{M}q)_1$ is a weak* closed face in $(p\mathcal{M}q)_1$. Since v is unique, we get that $v = pvq$ and so $F_{pq} \neq \emptyset$. Thus, F_{pq} contains extreme points by the Krein–Milman theorem. Since F_{pq} is a face, its extreme points are also extreme in $(p\mathcal{M}q)_1$ and so (1) follows from Lemma 3.10.

For (2), recall [37,1.4.8] that w is an extreme point of $(p\mathcal{M}q)_1$, if and only if

$$(p - ww^*)\mathcal{M}(q - w^*w) = \{0\}.$$

Fix a unitary u in $q\mathcal{M}q$. We have,

$$(p - vuu^*v^*)\mathcal{M}(q - u^*v^*vu) = (p - vv^*)\mathcal{M}u^*(q - v^*v)u$$
$$\subset [(p - vv^*)\mathcal{M}(q - v^*v)]u = \{0\},$$

and so vu is an extreme point of $(p\mathcal{M}q)_1$. $\quad\square$

THEOREM 3.12. (LYAPUNOV THEOREM FOR WEAKLY INFINITE VON NEU-MANN ALGEBRAS). *If $\mathcal{M}$ is a weakly infinite von Neumann algebra and Ψ is a norm continuous linear map of $\mathcal{M}$ into $\mathbb{C}^n$, then the following statements hold.*

- i) $\Psi(P)$ *is dense in* $\Psi((\mathcal{M}_+)_1)$.
- ii) $\Psi(S)$ *is dense in* $\Psi((\mathcal{M}_{sa})_1)$.
- iii) $\Psi(U)$ *is dense in* $\Psi((\mathcal{M})_1)$.
- iv) *If p and q are projections in $\mathcal{M}$ such that $q\mathcal{M}q$ is weakly infinite, then $\Psi(E((p\mathcal{M}q)_1))$ is dense in $\Psi((p\mathcal{M}q)_1)$.*

PROOF. As noted above, it is trivial that a von Neumann algebra is unital and has real rank zero. Hence, i), ii) and iii) are immediate corollaries of Theorem 3.6. To show that iv) holds, fix x in $(p\mathcal{M}q)_1$ and let $v|x|$ denote a polar decomposition of x. By (1) of Lemma 3.11, we may take v to be an extreme point of $p\mathcal{M}q$. Define Φ on $q\mathcal{M}q$ by

$$\Phi(y) = \Psi(vy).$$

It is clear that Φ is continuous and $q\mathcal{M}q$ is a weakly infinite von Neumann algebra by assumption. So, if we fix $\epsilon > 0$, by part iii) of the Theorem there is a unitary u in $q\mathcal{M}q$ with

$$\|\Phi(u - |x|)\| < \epsilon.$$

We have that vu is an extreme point of $(p\mathcal{M}q)_1$ by (2) of Lemma 3.11 and

$$\|\Psi(vu - x)\| = \|\Psi(v(u - |x|))\| = \|\Phi(u - |x|)\| < \epsilon;$$

hence, iv) holds, and the proof is complete. $\quad\square$

REMARKS. 1) The conclusion in part iv) of Theorem 3.12 does not hold for arbitrary weakly infinite C^*–algebras of real rank zero. Indeed, suppose p is a projection in $B(H)$ such that p and $1 - p$ each have infinite rank and let A denote the C^*–algebra generated by $pB(H)p, (1 - p)B(H)(1 - p)$, and the compact operators $\mathcal{K}$. If we put $q = 1 - p$, then $pAq = p\mathcal{K}q \subset \mathcal{K}$ and so $(pAq)_1$ does not contain any extreme points.

2) Note that Example 3.8 shows that a type 1 conclusion is not possible in Theorem 3.12. On the other hand, we show in section 6 that type 2 conclusions can hold even when the range of Ψ is allowed to be infinite dimensional. See Theorem 6.16.

§4. LYAPUNOV THEOREMS FOR
ATOMIC VON NEUMANN ALGEBRAS

In this section we shall consider the case where $\mathcal{M}$ is an **atomic** von Neumann algebra. We then have that $\mathcal{M}$ is the direct sum of type I factors, and so we may write

$$\mathcal{M} = \mathcal{M}_{fin} \oplus \mathcal{M}_{inf},$$

where $\mathcal{M}_{fin}$ is the direct sum of finite type I factors and $\mathcal{M}_{inf}$ is the direct sum of infinite type I factors. Since $\mathcal{M}_{inf}$ is weakly infinite, Theorem 3.12 shows that a Lyapunov Theorem of type 2 holds for $\mathcal{M}_{inf}$, even when Ψ is just norm continuous.

Now suppose $\mathcal{M}$ has a finite type I direct summand. By example 3.9, the best we can expect to get in this case is a Lyapunov of type 3. In the next theorem, which is the main result in this section, we show that a Lyapanov theorem of type 3 does hold for finite atomic von Neumann algebras. In order to state the theorem it is necessary to introduce some new notation. If $v = \langle \alpha_1, \ldots, \alpha_n \rangle$ is in $\mathbb{C}^n$, then we write

$$\|v\|_\infty = \max\{|\alpha_i| : i = 1, \ldots, n\} \qquad \text{and} \qquad \|v\|_1 = \sum_{i=1}^{n} |\alpha_i|$$

for the familiar ∞–norm and 1–norm. If Ψ maps an atomic von Neumann algebra $\mathcal{M}$ into $\mathbb{C}^n$, then we define

$$\delta(\Psi) = \delta = \sup\{\|\Psi(r)\|_1 : r \text{ is a minimal projection in } \mathcal{M}\}.$$

Also, for any projections p and q in M, put

$$\delta_{pq}(\Psi) = \delta_{pq} = \sup\{\|\Psi(x)\|_1 : x \text{ is of rank one in } (p\mathcal{M}q)_1\}.$$

Note that although we shall always use $\|\cdot\|_\infty$ to measure the distance between elements in the range of Ψ in this section, δ and δ_{pq} are defined using $\|\cdot\|_1$. Thus, it is possible that δ (resp. δ_{pq}) may be quite large, but since we are mainly interested in the case when δ is small, we believe this notation is justified. It should be emphasized that at this point δ is only defined for atomic von Neumann algebras. We shall give two natural extensions of this notion for arbtrary von Neumann algebras later (see Corollary 4.3 and Theorem 4.8).

The proof of the following theorem was inspired by Lindenstrauss's proof of Lyapunov's theorem [31] and a result due to Beck and Fiala [10]. (See Corollary 2.6 and Corollary 4.5 below).

THEOREM 4.1. (LYAPUNOV THEOREM FOR ATOMIC VON NEUMANN ALGE-
BRAS). *Suppose $\mathcal{M}$ is an atomic von Neumann algebra, Ψ is a weak* continuous
nonzero linear map of $\mathcal{M}$ into $\mathbb{C}^n$ and $\delta = \delta(\Psi)$ as defined above.*

  (i) *If Ψ is self–adjoint, then for each a in $(\mathcal{M}_+)_1$ there exists a projection
      p in $\mathcal{M}$ that commutes with a and such that $\|\Psi(p) - \Psi(a)\|_\infty < \delta$.*
 (ii) *If Ψ is self–adjoint, then for each b in $(\mathcal{M}_{sa})_1$ there exists a symmetry
      s in $\mathcal{M}$ that commutes with b and such that $\|\Psi(s) - \Psi(b)\|_\infty < 2\delta$.*
(iii) *For each normal element c in $(\mathcal{M})_1$ there exists a unitary element u in
      $\mathcal{M}$ that commutes with c and such that $\|\Psi(u) - \Psi(c)\|_\infty < 2\delta$.*
 (iv) *If p and q are projections in $\mathcal{M}$, $\delta_{pq} = \delta pq(\Psi)$ as defined above and
      $\Psi \neq 0$ on $p\mathcal{M}q$, then for each x in $(p\mathcal{M}q)_1$ there exists w in $E((p\mathcal{M}q)_1)$
      such that $\|\Psi(w) - \Psi(x)\|_\infty < 2\delta_{pq}$.*

PROOF. Note that $\delta > 0$ (resp. $\delta_{pq} > 0$) because $\mathcal{M}$ is atomic and $\Psi \neq 0$
(resp. $\Psi \neq 0$ on $p\mathcal{M}q$). We first prove parts ii) and iii) simultaneously. Let
$\mathcal{N}$ denote a MASA of $\mathcal{M}$ containing the normal element c (or the self–adjoint
element b for part (ii)). Write

$$T = \Psi^{-1}(\Psi(c)) \cap (\mathcal{N})_1$$

(for part (ii) replace T by $\Psi^{-1}(\Psi(b)) \cap (\mathcal{N}_{sa})_1$). Clearly T is convex and, since Ψ
is weak* continuous, T is compact (in both cases). Hence, by the Krein–Milman
theorem [39.p.70] T contains an extreme point e, which is self–adjoint in part
(ii). Write r for the spectral projection of e corresponding to $\{z : |z| < 1\}$,
and, for each natural number $m > 1$, let r_m denote the spectral projection of
e corresponding to $\{z : |z| < 1 - 1/m\}$. We claim that Ψ is injective on each
$r_m\mathcal{N}$. Indeed, if this is not the case, then there exists $d \neq 0$ in the intersection
of the kernel of Ψ and $r_m\mathcal{N}$. (For part (ii), since Ψ is self–adjoint, d can be
chosen to be self–adjoint.) Multiplying by a small positive number if necessary,
we may assume d has norm less than $1/m$. In this case both $e + d$ and $e - d$ lie
in T, contradicting the extremity of e. Hence Ψ is injective on each $r_m\mathcal{N}$, and
therefore each $r_m\mathcal{N}$ has dimension at most n. Since $r_m \nearrow r$ in $\mathcal{M}$, $r = r_m$ for
all large m. Thus, $r\mathcal{N}$ has dimension $k \leq n$.

Next observe that, if $r = 0$, then e is unitary (and self–adjoint in part (ii)).
Since $\Psi(e) = c$ (or $\Psi(e) = b$ in part (ii)) by its definition, the conclusions of
parts (ii) and (iii) are true for all n in this case. Also e commutes with c (resp.
b) since it lies in $\mathcal{N}$.

So suppose $r \neq 0$. To complete the proof of parts (ii) and (iii), we proceed
by induction on the dimension n of the range of Ψ. First assume $n = 1$ and let
Ψ_0 denote the restriction of Ψ to $r\mathcal{N}$. In this case Ψ_0 is given by a 1×1 matrix
and, since r is minimal, we have

$$\|\Psi_0(r)\|_1 = \|\Psi(r)\|_1 \leq \delta.$$

Write $u = r + (1 - r)e$ and note that by our definition of r, u is unitary (and
self–adjoint if e is self–adjoint). Also, since r has rank one we have that $re = \alpha r$
for some complex scalar α and $|\alpha| < 1$ by our definition of r. Thus,

$$\|\Psi(u - c)\|_\infty = \|\Psi(u - e)\|_\infty = \|\Psi(r(1 - e))\|_\infty = \|\Psi((1 - \alpha)r)\|_\infty$$
$$= |1 - \alpha| \|\Psi_0(r)\|_\infty = |1 - \alpha| \|\Psi_0(r)\|_1 < 2\delta.$$

Similarly, we have $\|\Psi(u - b)\|_\infty < 2\delta$ in the self–adjoint case. Hence, (ii) and (iii) are true if $n = 1$.

Now suppose that $n > 1$ and parts (ii) and (iii) of the theorem are true for any Ψ whose range lies in $\mathbb{C}^j$ with $1 \leq j < n$. Since $r\mathcal{N}$ is finite dimensional, there are projections $q_1, \ldots, q_k$ in $\mathcal{N}$ that are minimal in $\mathcal{N}$ and form a basis for $r\mathcal{N}$. We have that $q_1 + \cdots + q_k = r$ and every element of $r\mathcal{N}$ has the form $\sum \alpha_i q_i$. Let Ψ_0 denote the restriction of Ψ to $r\mathcal{N}$ and consider the matrix for Ψ_0 determined by the basis $\{q_1, \ldots, q_k\}$ in $r\mathcal{N}$ and the standard basis for $\mathbb{C}^n$. The ith column of this matrix is the vector $\Psi(q_i)$. Since the q_i's are minimal in $\mathcal{N}$ and $\mathcal{N}$ is a MASA in $\mathcal{M}$, the q_i's are also minimal in $\mathcal{M}$. Hence, the ℓ^1 norm of each column is no more than δ. Since $k \leq n$, there are no more columns than rows in the matrix of Ψ_0. Hence, at least one of the n rows of the matrix has ℓ^1 norm no greater than δ. Permuting rows if necessary, we may assume that the ℓ^1 norm of the nth row is no greater than δ.

By deleting the nth row of the matrix of Ψ_0, we get the matrix for a new operator Ψ_1 that maps $r\mathcal{N}$ into $\mathbb{C}^{n-1}$. Clearly we have

$$\delta(\Psi_1) \leq \delta(\Psi).$$

Since the range of Ψ_1 lies in $\mathbb{C}^{n-1}$, we may apply our induction hypothesis to Ψ_1 and get a unitary u_0 (self–adjoint in part (ii)) in $r\mathcal{N}$ such that

$$(*) \qquad\qquad \|\Psi_1(u_0 - re)\|_\infty < 2\delta(\Psi_1) \leq 2\delta.$$

Set

$$u = u_0 + (1 - r)e$$

(which is self–adjoint if Ψ is self–adjoint). By the definition of r, u is unitary. Note that

$$\Psi(u - c) = \Psi(u - e) = \Psi(u_0 - re)$$

(or in the self–adjoint case $\Psi(u - b) = \Psi(u_0 - re)$). Since $u_0 - re$ lies in $r\mathcal{N}$,

$$\|\Psi(u_o - re)\|_\infty = \|\Psi_0(u_0 - re)\|_\infty$$
$$= \max\{\|\Psi_1(u_0 - re)\|_\infty, \; |\varphi[\Psi_0(u_0 - re)]|\},$$

where φ is the linear functional on $\mathbb{C}^n$ corresponding to evaluation of the nth coordinate. By $(*)$ above, we have $\|\Psi_1(u_0 - re)\|_\infty < 2\delta$. Also, since $r\mathcal{N}$ is finite–dimensional and the spectrum of re is contained in $(0, 1)$, we have $\|re\| < 1$. Hence,

$$\|u_0 - re\|_\infty < 2.$$

Because the last row of the matrix for Ψ_0 has ℓ^1 norm no greater than δ, we get $|\varphi[\Psi_0(u_0 - re)]| < 2\delta$. Therefore,

$$\|\Psi(u - c)\|_\infty < 2\delta$$

(or $\|\{\Psi(u - b)\|_\infty < 2\delta$ in the self–adjoint case), and the proof of parts (ii) and (iii) of the Theorem is complete.

To prove part (i), put $b = 2a - 1$ and note that b is in $(\mathcal{M}_{sa})_1$. By part (ii) of the Theorem there exists symmetry s that commutes with b and such that $\|\Psi(s) - \Psi(b)\|_\infty < 2\delta$. If we write

$$p = (s + 1)/2,$$

then p is a projection, and

$$\Psi(p) - \Psi(a) = \frac{1}{2}\Psi(s + 1) - \frac{1}{2}\Psi(b + 1) = \frac{1}{2}(\Psi(s) - \Psi(b)).$$

Since $\|\Psi(s) - \Psi(b)\|_\infty < 2\delta$, $\|\Psi(p) - \Psi(a)\|_\infty < \delta$, and the proof of part (i) is complete.

To show that (iv) holds, we proceed as in the proof of part iv) of Theorem 3.12. Let $v|x|$ denote a polar decomposition of x. By part (1) of Lemma 3.11, v may be chosen to be an extreme point of $(p\mathcal{M}q)_1$. Define $\Phi : q\mathcal{M}q \to \mathbb{C}^n$ by

$$\Phi(y) = \Psi(vy).$$

By part (iii) of this Theorem, there is a unitary operator u in $q\mathcal{M}q$ such that

$$\|\Phi(u - |x|)\|_\infty < 2\delta(\Phi).$$

Since vx is of rank one for each minimal projection x in $q\mathcal{M}q$, we have

$$\delta(\Phi) \leq \delta_{pq}.$$

Hence, as in the proof of 3.12,

$$\|\Psi(vu - x)\|_\infty = \|\Phi(u - |x|)\|_\infty < 2\delta_{pq}.$$

Finally, we have that $w = vu \in E((p\mathcal{M}q)_1)$ by part (2) of Lemma 3.11. $\square$

REMARK. Note that it is possible that $u = -1$ in part (ii) of Theorem 4.1. In this case we would have that $p = (u + 1)/2 = 0$ in part (i). Similarly, if $u = 1$ in part (ii), then $p = 1$.

COROLLARY 4.2. *Part (i) of Theorem 4.1 holds for arbitrary Ψ if we replace δ by 2δ. Similarly, part (ii) holds for arbitrary Ψ with 2δ replaced by 4δ.*

PROOF. Suppose Ψ is a not necessarily self-adjoint weak* continuous map of $\mathcal{M}$ into $\mathbb{C}^n$. Then Ψ has the form

$$\Psi(x) = (f_1(x), \ldots, f_n(x)),$$

where the f_j's are weak* continuous linear functionals. Define a map Φ from $\mathcal{M}$ into $\mathbb{C}^{2n}$ by

$$\Phi(x) = (\mathrm{Re}(f_1(x)), \, \mathrm{Im}(f_1(x)) \ldots, \, \mathrm{Re}(f_n(x)), \, \mathrm{Im}(f_n(x))).$$

Also, define Λ on $\mathbb{C}^{2n}$ by

$$\Lambda(\alpha_1, \alpha_2, \ldots, \alpha_{2n-1}, \alpha_{2n}) = (\alpha_1 + i\alpha_2, \ldots, \alpha_{2n-1} + i\alpha_{2n}).$$

Clearly, we have $\Psi = \Lambda \circ \Phi$. Observe that $\Lambda\Lambda^* = 2(1_n)$, and so $\|\Lambda\| = \sqrt{2}$. Also, if $x \in \mathcal{M}$, then

$$|\text{Re}(f_i(x))| + |\text{Im}(f_i(x))| \leq \sqrt{2}|f_i(x)|,$$

and so

$$\delta(\Phi) \leq \sqrt{2}\delta(\Psi).$$

If $a \in (\mathcal{M}_+)_1$, then we may apply part i) of Theorem 4.1 to find a projection p such that $\|\Phi(p - a)\| \leq \delta(\Phi)$, and therefore

$$\|\Psi(p - a)\| = \|\Lambda \circ \Phi(p - a)\| \leq \|\Lambda\| \, \|\Phi(p - a)\| \leq 2\delta(\Psi).$$

Thus, our first assertion holds. The proof of the second assertion is similar. $\square$

REMARK. A version of Theorem 4.1 also holds in the more general case where Ψ is an affine map of a weak* closed face F of $(\mathcal{M})_1$. We now briefly sketch this generalization in the case where F is a positive face. Recall (Theorem 2.1, part (1)) that in this case we have $F = [r, s]$ for some projections $r < s$. We write

$$\delta(F, \Psi) = \sup\{\|\Psi(q) - \Psi(r)\|_1 : q - r \text{ has rank one}\}.$$

Our new version of part i) of Theorem 4.1 may be stated as follows:

> If $\mathcal{M}$ is atomic, Ψ is a non–constant, self-adjoint, affine weak* continuous map from the weak* closed face F of $(\mathcal{M}_+)_1$ into $\mathbb{C}^n$, then for each a in F there is a projection p in F that commutes with a such that $\|\Psi(p) - \Psi(a)\| < \delta(F, \Psi)$.

For the proof, it is sufficient to show that we may reduce to the situation of Theorem 4.1, part 1. To do this, note that by part 1) of Proposition 2.2, we may assume that Ψ is an affine map of $(\mathcal{N}_+)_1$ for some von Neumann subalgebra $\mathcal{N}$. In fact 2.2 gives us a weak* continuous map Ψ on the order interval $[0, 1]$ of a weak* closed hereditary subalgebra of $\mathcal{M}$. So, we are reduced to showing that such a map has a unique weak* continuous linear extension to all of $\mathcal{N}$. Using Proposition 1.5 we may extend Ψ to a norm continuous linear map on $\mathcal{N}$. We claim that this extension is weak* continuous. Indeed, if $\{p_\alpha\}$ is an increasing net of projections in $\mathcal{N}$ with limit p, then $\lim \Psi(p_\alpha) = \Psi(p)$ because Ψ is weak* continuous on the order interval $[0, 1]$. Thus for any linear functional f on $\mathbb{C}^n$, $\Psi^*(f)$ is normal; hence, Ψ is weak* continuous [42, p.136]. The reduction is complete.

The generalizations in the other 3 cases are similar.

In the next corollary we combine our special Lyapunov theorems into one general result. Before stating it, we must extend our definition of δ so that it applies to an arbitrary von Neumann algebra $\mathcal{M}$. Recall from the introduction that we have the decomposition

$$\mathcal{M} = \mathcal{M}_{na} \oplus \mathcal{M}_{inf} \oplus \mathcal{M}_{fin}.$$

If Ψ maps $\mathcal{M}$ into $\mathbb{C}^n$, then we write

$$\delta_{fin}(\Psi) = sup\{\|\Psi(r)\|_1 : r \text{ is a minimal projection in } \mathcal{M}_{fin})\}.$$

Also, if p and q are projections in $\mathcal{M}$, then put

$$\delta_{fin}^{pq}(\Psi) = sup\{\|\Psi(x)\|_1 : x \text{ has rank one in } p(\mathcal{M}_{fin})q\}.$$

If $\mathcal{M}_{fin} = \{0\}$ (resp. $p(\mathcal{M}_{fin})q = \{0\}$), then we put $\delta_{fin} = 0$ (resp. $\delta_{fin}^{pq} = 0$).

COROLLARY 4.3 (LYAPUNOV THEOREM FOR VON NEUMANN ALGEBRAS). *If Ψ is a weak* continuous map of a von Neumann algebra $\mathcal{M}$ into $\mathbb{C}^n$ and*

$$\delta_{fin}(\Psi) > 0,$$

then the following statements are true.
  (i) *If Ψ is self–adjoint, then for each a in $(\mathcal{M}_+)_1$ there exists a projection p in $\mathcal{M}$ such that $\|\Psi(p) - \Psi(a)\|_\infty < \delta_{fin}$.*
  (ii) *If Ψ is self–adjoint, then for each b in $(\mathcal{M}_{sa})_1$ there exists a symmetry s in $\mathcal{M}$ such that $\|\Psi(s) - \Psi(b)\|_\infty < 2\delta_{fin}$.*
  (iii) *For each normal element c in $(\mathcal{M})_1$ there exists a unitary element u in M such that $\|\Psi(u) - \Psi(c)\|_\infty < 2\delta_{fin}$.*
  (iv) *If p and q are projections in M and $\delta_{fin}^{pq} > 0$, then for each x in $(p\mathcal{M}q)_1$ there exists w in $E((p\mathcal{M}q)_1)$ such that $\|\Psi(w) - \Psi(x)\|_\infty < 2\delta_{fin}^{pq}$.*

PROOF. We shall only prove part (i). The proof in the other cases is similar. If we fix a in $(\mathcal{M}_+)_1$, then we have

$$a = a_{na} \oplus a_{inf} \oplus a_{fin},$$

where $a_{na} \in \mathcal{M}_{na}, a_{inf} \in \mathcal{M}_{inf}$ and $a_{fin} \in \mathcal{M}_{fin}$. If Ψ is self–adjoint, then we may use part i) of Theorem 4.1 to find a projection p_{fin} in $\mathcal{M}_{fin}$ such that

$$\|\Psi(p_{fin} - a_{fin})\| < \delta_{fin}.$$

Using part i) of Theorems 2.5 and 3.12 we may find projections $p_{na} \in \mathcal{M}_{na}$ and $p_{inf} \in \mathcal{M}_{inf}$, respectively such that $\|\Psi(p_{na} \oplus p_{inf} - a_{na} \oplus a_{inf})\|$ is so small that with $p = p_{na} \oplus p_{inf} \oplus p_{fin}$ we have

$$\|\Psi(p - a)\| < \delta_{fin}.$$

Hence, the statement in part (i) holds. $\square$

Note that in Corollary 4.3 p need not commute with a.

In [23 Theorem 2.1,p.298] Elton and Hill prove a theorem which is superficially similar to a special case of Corollary 4.3(i). We restate it in the language of the present paper.

> If $\mathcal{M}$ is an abelian von Neumann algebra and $\Psi : \mathcal{M} \to \mathbb{C}^n$ is a weak* continuous, positive linear map, then for every $a \in (\mathcal{M}_+)_1$ there is a projection $p \in \mathcal{M}$ such that $\|\Psi(a - p)\|_\infty \leq \frac{\alpha n}{2}$, where
>
> $$\alpha = sup\{\|\Psi(r)\|_\infty : r \text{ is a minimal projection in } \mathcal{M}\}.$$

This result differs from Corollary 4.3 in two crucial ways. First, Elton and Hill use $\|\cdot\|_\infty$ in their definition of α, while we use $\|\cdot\|_1$ in our definition of δ. Both δ and α measure the "size" of the atoms, but they use different norms to measure this quantity. Second, the Elton and Hill estimate depends on the dimension n of the range of Ψ, while our results do not. Since we have in mind noncommutative and infinite dimensional generalizations of Corollary 4.3, it seems more natural to us to avoid any dependence on the dimension of the range of Ψ.

The next Corollary shows what Theorem 4.1 implies for finite matrices. We write $\ell^\infty(m)$ for $(\mathbb{C}^m, \|\cdot\|_\infty)$.

COROLLARY 4.4. *Suppose n and m are positive integers, $a = \{\alpha_{ij}\}$ is a nonzero $n \times m$ complex matrix and*

$$\delta = \max_j \left\{ \sum_{i=1}^n |\alpha_{ij}| \right\}.$$

   i) *If $v \in ((\ell^\infty(m))_+)_1$, then there is an m–dimensional vector p whose entries are 0's and 1's such that $\|a(p - v)\|_\infty < \delta$.*
   ii) *If $v \in (\ell^\infty(m)_{sa})_1$, then there is an m–dimensional vector s whose entries are ± 1 such that $\|a(s - v)\|_\infty < 2\delta$.*
   iii) *If $v \in (\ell^\infty(m))_1$, then there is an m–dimensional vector u whose entries have modulus 1 such that $\|a(u - v)\|_\infty < 2\delta$.*

PROOF. Define $\Psi : \ell^\infty(m) \to \mathbb{C}^n$ by $\Psi(x) = ax$. Since the dimension of the domain of Ψ is finite, Ψ is weak* continuous and Theorem 4.1 applies. Note that if q is a minimal projection in $\ell^\infty(m)$, then the coordinates of q are 0 except for one index, where the value is 1. If we let j denote the index of the non–zero entry of q, then we have

$$\|\Psi(q)\|_1 = \sum_{i=1}^n |\alpha_{ij}|$$

and so $\delta(\Psi) = \delta$. The desired conclusions now follow from parts i), ii) and iii) of Theorem 4.1. $\square$

Before stating the next Corollary it is convenient to introduce some notation. Suppose X is a set and $\mathcal{E}$ is a collection of subsets of X. For x in X, the **degree** $d(x)$ of x is, by definition, the cardinality of $\{E \in \mathcal{E} : x \in E\}$ and the **maximum degree** of $(X, \mathcal{E})$ is $\max\{d(x) : x \in X\}$.

LYAPUNOV THEOREMS FOR OPERATOR ALGEBRAS   37

COROLLARY 4.5: (BECK-FIALA [10]). *If* $X = \{x_1, \ldots, x_m\}$ *is a set,* $\mathcal{E} = \{E_1, \ldots, E_n\}$ *is a nonvoid collection of nonvoid subsets of* X *and* w *is a real–valued function on* X, *then there are integers* $k_1, \ldots, k_n$ *such that* $|k_j - w(x_j)| < 1$ *for each* j *and*

$$\left| \sum_{x_j \in E_i} k_j - \sum_{x_j \in E_i} w(x_j) \right| < d, \qquad i = 1, \ldots, n,$$

where d *denotes the maximum degree of* $(X, \mathcal{E})$.

PROOF. Replacing $w(x_j)$ by $w(x_j) - [w(x_j)]$, if necessary, we may assume $0 \leq w(x_j) \leq 1$. (here $[\cdot]$ denotes the greatest integer function.) Let a denote the incidence matrix determined by $(X, \mathcal{E})$. Thus, a is the n by m matrix whose (i, j)–entry is 1 if $x_j \in E_i$ and 0, otherwise. Since a is not the zero matrix, Corollary 4.4 applies. For each fixed j we have

$$\sum_{i=1}^{n} |\alpha_{ij}| = d(x_j) \leq d,$$

where α_{ij} denotes the (i, j)–entry of a. Hence, if we write

$$w = \langle w(x_1), \ldots, w(x_m) \rangle,$$

then there is a vector $e = \langle \epsilon_1, \ldots, \epsilon_m \rangle$, where each ϵ_i is either 0 or 1 such that $\|a(e - w)\|_\infty < d$. Setting $k_j = \epsilon_j$ gives the desired result. $\square$

It is natural at this point to ask if similar conclusions hold if other standard norms are used on the range or in the definition of δ. The next two examples provide some negative evidence.

EXAMPLE 4.6. In this example we show that if we replace $\| \cdot \|_\infty$ as the norm on the range $\mathbb{C}^n$ with $\| \cdot \|_p$ ($1 \leq p < \infty$) then the conclusion in part i) of Theorem 4.1 no longer holds. Fix $1 \leq p < \infty$ and a positive integer n. Define $\Psi : \ell^\infty(n) \to \ell^p(n)$ by

$$\Psi(\langle \lambda_1, \ldots, \lambda_n \rangle) = \langle \lambda_1, \ldots, \lambda_j, 0, \ldots, 0 \rangle.$$

In this case $\delta = 1$. However, for any projection q in $\ell^\infty(n)$ we have

$$\|\Psi(q - (1/2)1)\|_p \geq \frac{j^{1/p}}{2}.$$

Thus, the conclusion in part i) of Theorem 4.1 is false in this case.

The next example shows that things can go wrong if the constant δ is defined using an ℓ^p–norm with $p > 2$. It is an adaptation of an idea of Spencer [41,p.703]. Recall that an n by n matrix H is said to be a Hadamard matrix if its entries are 1 or -1 and HH^* is n times the identity matrix. This example shows that the dependence of any approximation on the dimension of the range, such as found in Elton and Hill [23, Th. 1.2], cannot be avoided when δ is defined using an ℓ^p norm for $p > 2$. The case $p = 2$ leads to the Komlos Conjecture [41,p. 698].

EXAMPLE 4.7. Let Ψ be given by a Hadamard matrix mapping $\ell^\infty(n)$ onto $\ell^\infty(n)$. If q is a minimal projection in $\ell^\infty(n)$, then $\|\Psi(q)\|_p = n^{1/p}$, and so $\delta = n^{1/p}$. If u is a vector in $\ell^\infty(n)$ whose entries are 1 or -1 (i.e., if u is a self–adjoint extreme point in Q), then we claim that $\|\Psi(u)\|_\infty \geq \sqrt{n}$. Indeed, since $\|u\|_2 = \sqrt{n}$, we have that $\|\Psi(u)\|_2 = n$. Thus, if $\Psi(u) = (a_1, \ldots, a_n)$, then $\Sigma|a_j|^2 = n^2$, so for some value of j, $|a_j|^2 \geq n$, and therefore $\|\Psi(u)\|_\infty \geq \sqrt{n}$. Hence,

$$\frac{\|\Psi(u)\|_\infty}{\delta} \geq n^{(1/2 - 1/p)},$$

which is unbounded as n increases (and $p > 2$). Thus, the conclusion in part ii) of Theorem 4.1 is false in this case.

If we drop the assumption that $\Psi(\mathcal{M}) \subset \mathbb{C}^n$ and merely require that the range lie in $\ell^\infty(\mathbb{N})$ then, as shown in Example 2.7, it may be that Ψ is an isometry. Note, however, that if q is a minimal projection in $\mathcal{M}_{fin}$ then, in the notation of Example 2.7, there is a state f in the predual $\mathcal{M}_*$ with $f(q) = 1$. Since the f_j's are dense in $\mathcal{M}_*$, $\{j : f_j(q) > 1/4\}$ is infinite. Hence, $\|\Psi(q)\|_1 = \infty$, and so, if we define $\delta_{fin}(\Psi)$ as above, then $\delta_{fin}(\Psi) = \infty$. On the other hand, if $\mathcal{M}$ is nonatomic, then $\delta_{fin} = 0$. Yet by Example 2.7, Ψ may map extreme points to extreme points. This suggests that our definition of δ_{fin} may not be the appropriate notion when the range of Ψ is allowed to lie in $\ell^\infty(\mathbb{N})$. In fact another extension of the definition of our original δ to general von Neumann algebras is available.

Write

$$\delta^\infty(\Psi) = \sup\bigl\{ \inf\{ \|\Psi(q)\|_1 : q \leq r \text{ and } q \in P_\mathcal{M} \} : r \in P_\mathcal{M} \bigr\}.$$

Also, if p and q are projections in $\mathcal{M}$, then we put

$$\delta_{pq}^\infty(\Psi) = \sup\bigl\{ \inf\{ \|\Psi(x)\|_1 : x^*x < r \text{ and } \|x\| = 1 \} : r \in P_\mathcal{M} \bigr\}.$$

It is clear that if $\mathcal{M}$ is a atomic von Neumann algebra, then $\delta^\infty = \delta$. On the other hand, when $\mathcal{M}$ is nonatomic, we have $\delta_{fin}(\Psi) = 0$ for any Ψ (even when the range of Ψ has infinite dimension); but, when Ψ is defined as in Example 2.7, we have $\delta^\infty(\Psi) = \infty$.

In fact, if $\mathcal{M}$ is nonatomic and Ψ is a positive map of $\mathcal{M}$ into $\ell^\infty(\mathbb{N})$, then $\delta^\infty(\Psi)$ is either 0 or ∞. Indeed, if there is a nonzero projection $p \in \mathcal{M}$ such that $\|\Psi(p)\|_1 < \infty$, then, since $\mathcal{M}$ is nonatomic, for any positive integer n, we may find nonzero projections $q_1, \ldots, q_n$ such that $\sum_{i=1}^n q_i = p$. Since Ψ is positive, we have $\|\Psi(p)\|_1 = \sum_{i=1}^n \|\Psi(q_i)\|_1$ and so $\|\Psi(q_i)\|_1 \leq \frac{1}{n}\|\Psi(p)\|_1$ for at least one i. Thus, $\delta^\infty(\Psi) = 0$ in this case.

THEOREM 4.8. *If $\mathcal{M}$ is a von Neumann algebra, Ψ is a weak* to weak* continuous linear map of $\mathcal{M}$ into $\ell^\infty(\mathbb{N})$ and $\epsilon > 0$, then the following statements hold.*

    i) *If Ψ is self–adjoint and $a \in (\mathcal{M}_+)_1$, then there is a projection p in $\mathcal{M}$ such that $\|\Psi(p - a)\|_\infty \leq \delta^\infty + \epsilon$.*

ii) If Ψ is self–adjoint and $b \in (\mathcal{M}_{sa})_1$, then there is a symmetry s in $\mathcal{M}$ such that $\|\Psi(s - b)\|_\infty \leq 2\delta^\infty + \epsilon$.
iii) If c is a normal element of $(\mathcal{M})_1$, then there is a unitary u in $\mathcal{M}$ such that $\|\Psi(u - c)\|_\infty \leq 2\delta^\infty + \epsilon$.
iv) If p and q are projections in $\mathcal{M}$ and $x \in (p\mathcal{M}q)_1$, then there is an element w in $E((p\mathcal{M}q)_1)$ such that $\|\Psi(w - x)\|_\infty \leq 2\delta^\infty_{pq} + \epsilon$.

Moreover if $\mathcal{M}$ is atomic and abelian then each of i), ii), iii) and iv) hold with $\epsilon = 0$.

PROOF. First assume that $\mathcal{M}$ is atomic and abelian. In this case the set of projections in $\mathcal{M}$ is weak* compact. For each natural number n define Ψ_n on $\mathcal{M}$ by $\Psi_n = \pi_n \circ \Psi$, where π_n is the projection of ℓ^∞ onto its first n coordinates. To show that i) is true, for each n write

$$G_n = \{p \in \mathcal{M} : p \text{ is a projection and } \|\Psi_n(p - a)\|_\infty \leq \delta^\infty\}.$$

Each G_n is weak* compact in $(\mathcal{M}_+)_1$. Since $\mathcal{M}$ is atomic, we have $\delta^\infty = \delta$ and we may apply part i) of Theorem 4.1 to conclude that each G_n is nonvoid; also the elements of $\{G_n : n = 1, \dots\}$ are ordered by reverse inclusion. Thus there is a projection p in the intersection of all the G_n's. Clearly

$$\lim_{n \to \infty} \Psi_n(p - a) = \Psi(p - a)$$

in the weak* topology. Since $\|\Psi_n(p - c)\|_\infty \leq \delta^\infty$ for each n, and since the norm is weak* lower semicontinuous, we get $\|\Psi(p - c)\|_\infty \leq \delta^\infty$, as desired.

Next observe that the sets of symmetries and unitaries in $\mathcal{M}$ are also weak* closed. In addition, $E((p\mathcal{M}q)_1) = pqE((\mathcal{M})_1)$ and so it too is weak* closed. Hence, we may argue in a similar fashion to show that ii) and iii) hold iv) when $\mathcal{M}$ is atomic and abelian.

For part i) in the general case we may use spectral theory to find commuting projections $\{p_1, \dots, p_n\}$ and complex scalars $\{\lambda_1, \dots, \lambda_n\}$ such that

$$(*) \qquad \left\| a - \sum_{i=1}^{n} \lambda_i p_1 \right\| < \frac{\epsilon}{2\|\Psi\|}.$$

By a simple argument using Zorn's Lemma and the definition of δ^∞, for each i, we may find a family $\{q_{\alpha i}\}$ of commuting projections such that $\sum_\alpha q_{\alpha i} = p_i$ and

$$\|\Psi(q_{\alpha i})\| < \delta^\infty(\Psi) + \frac{\epsilon}{2}.$$

Let $\mathcal{N}$ denote a MASA in $\mathcal{M}$ that contains all of the $q_{\alpha i}$'s and note that we have $\delta^\infty(\Psi|_\mathcal{N}) \leq \delta^\infty(\Psi) + \frac{\epsilon}{2}$. Since $\mathcal{N}$ is atomic and abelian, by the first part of the proof, there is a projection p such that

$$(**) \qquad \left\| \Psi(p) - \Psi(\sum_{i=1}^{n} \lambda_i p_i) \right\| \leq \delta^\infty(\Psi) + \frac{\epsilon}{2}.$$

Using (∗) and (∗∗) above, we get $\|\Psi(p) - \Psi(a)\| \leq \delta^\infty + \epsilon$, as desired.

A similar argument shows that ii) and iii) also hold in the general case.

Finally for iv) we argue as in the proofs of the fourth parts of Theorem 3.12 and 4.1. Fixing a polar decomposition $v|x|$ for x, with v taken to be an extreme point, and defining Φ as before, we have $\delta^\infty(\Phi) \leq \delta^\infty(\Psi)$. As before, we may use part iii) of the the Theorem to find a unitary u in $q\mathcal{M}q$ such that

$$\|\Psi(vu - x)\| \leq \delta^\infty(\Psi) + \epsilon.$$

By part (2) of Lemma 3.11, we have $w = vu \in E((p\mathcal{M}q)_1)$. □

CONJECTURE 4.9. *If $\mathcal{M}_{fin} = \{0\}$, then the conclusions of Theorem 4.8 hold with $\epsilon = 0$.*

§5. SIMULTANEOUS APPROXIMATIONS

In this section $\mathcal{N}$ **shall denote an abelian von Neumann algegba.** We continue to let Ψ stand for a weak* continuous self–adjoint linear map of $\mathcal{N}$ into the finite–dimensional space $\mathbb{C}^n$. Note that since $\mathcal{N}$ is abelian, it has no infinite atomic part, and so we may define z_{fin} and z_{na} as in the introduction and write

$$\mathcal{N} = \mathcal{N}_{fin} \oplus \mathcal{N}_{na}.$$

Thus, with $\delta = \delta(\Psi)$ defined as in Section 4, the conclusions of Theorem 4.1 hold for Ψ applied to $\mathcal{N}_{fin}$, and the conclusions of Theorem 2.5 hold for the restriction of Ψ to $\mathcal{N}_{na}$.

We shall show that simultaneous versions of Theorem 2.5 and Theorem 4.1 hold for $\mathcal{N}_{na}$ and $\mathcal{N}_{fin}$. In the former case, we show (Theorem 5.3) that, given a partition of unity in $\mathcal{N}_{na}$, i.e. $\{a_1, \ldots, a_k\}$ in $((\mathcal{N}_{na})_+)_1$ with $\sum a_j = 1$, then there are orthogonal projections $\{p_1, \ldots, p_k\}$ such that $\sum p_j = 1$ and $\Psi(p_j - a_j) = 0$. In the latter case, we prove a simultaneous "type 3" result (Theorem 5.6), but we must restrict ourselves to the case where each $a_i = (1/k)1$.

Let us begin by introducing some additional notation. Fix an integer $k > 1$ and let $\widetilde{\mathcal{N}}$ denote the direct sum of k copies of $\mathcal{N}$. Write

$$Q = \left\{ a \in (\widetilde{\mathcal{N}}_+)_1 : a = a_1 \oplus \cdots \oplus a_k \text{ and } \sum_{j=1}^{k} a_j = 1 \right\}$$

Clearly, Q is convex and weak* closed.

LEMMA 5.1. *If $\mathcal{N}$ is abelian and nonatomic, then*

i) *$p = p_1 \oplus \cdots \oplus p_k$ is an extreme point of Q if and only if each p_j is a projection.*

ii) *The facial dimension of Q is infinite.*

PROOF. Suppose $a_1 \oplus \cdots \oplus a_k$ denotes an element of Q such that at least one a_j is not a projection. We will show that this element is composite in Q. Relabeling if necessary, we may assume that a_1 is not a projection. In this case there is $\delta > 0$ such that, if q denotes the spectral projection of a_1 corresponding to $(\delta, 1 - \delta)$, then $q \neq 0$. We have

$$\delta q < a_1 q < (1 - \delta)q$$

and therefore we also have

$$\delta q < (1 - a_1)q < (1 - \delta)q.$$

Since $a_2 + \cdots + a_k = 1 - a_1$, there is some $j > 1$ such that $a_j q \neq 0$ and $a_j q$ is not a projection. Relabeling again, we may assume $j = 2$. Since $a_2 q$ is not a projection, by decreasing δ if necessary we may find a spectral projection r of $a_2 q$ such that

$$\delta r < a_2 r < (1 - \delta) r.$$

Now fix a nonzero, self-adjoint element b in $\mathcal{N}$ such that $-r \leq b \leq r$. Since $r \leq q$,

$$(a_1 + \delta b) \oplus (a_2 - \delta b) \oplus \cdots \oplus a_k \text{ and } (a_1 - \delta b) \oplus (a_2 + \delta b) \oplus \cdots \oplus a_k$$

both lie in Q. Hence, $a_1 \oplus \cdots \oplus a_k$ is a composite point in Q. Thus, if $p_1 \oplus \cdots \oplus p_k$ is an extreme point in Q, then each p_j is a projection. Conversely, if p is a projection in Q, then p is certainly an extreme point. Hence, i) holds.

To see that ii) is true, fix a composite point $a_1 \oplus \cdots \oplus a_k$ in Q. By the first paragraph of the proof there is a projection r in $\mathcal{N}$ such that if $-r \leq b \leq r$ then

$$seg(a_1 \oplus \cdots \oplus a_k + \delta(b \oplus (-b) \oplus 0 \oplus \cdots \oplus 0), a_1 \oplus \cdots \oplus a_k - \delta(b \oplus (-b) \oplus 0 \oplus \cdots \oplus 0)) \subset Q.$$

Thus, if we put

$$\mathcal{Y} = \mathbb{R}\text{–span}(\{b \oplus (-b) \oplus 0 \cdots \oplus 0 : -r \leq b \leq r\}),$$

then $\mathcal{Y}$ is lineal with respect to $a_1 \oplus \cdots \oplus a_k$ and $\mathcal{Y}$ is infinite-dimensional because $\mathcal{N}$ is nonatomic. Hence, the facial dimension of Q is infinite by Proposition 1.4. $\square$

As the next example shows, Lemma 5.1 does not generalize to the noncommutative case.

EXAMPLE 5.2. Write

$$p_1 = \frac{1}{2}\begin{bmatrix} 1 & 1 \\ 1 & 1 \end{bmatrix}, \ p_2 = \frac{1}{2}\begin{bmatrix} 1 & \omega \\ \omega^* & 1 \end{bmatrix} \text{ and } p_3 = \frac{1}{2}\begin{bmatrix} 1 & \omega^* \\ \omega & 1 \end{bmatrix},$$

where $\omega = -\frac{1}{2} + i(\frac{\sqrt{3}}{2})$, and note that each p_i is a projection. Define

$$a = \frac{2}{3}(p_1 \oplus p_2 \oplus p_3).$$

If a is a composite point, then we may find $b = b_1 \oplus b_2 \oplus b_3$ and $c = c_1 \oplus c_2 \oplus c_3$ such that $a \neq b$, $a \neq c$, $\sum b_i = \sum c_i = 1$, all terms lie between 0 and 1 and $a = (\frac{1}{2})(b + c)$. If we write

$$b_i = a_i + y_i,$$

where $a_i = (\frac{2}{3})p_i$, then it follows that

$$c_i = a_i - y_i.$$

We have

$$(1 - p_i)a_i(1 - p_i) = (\frac{1}{2})(1 - p_i)(b_i + c_i)(1 - p_i) = 0,$$

and, since all terms are positive, we get

$$(1 - p_i)b_i(1 - p_i) = (1 - p_i)c_i(1 - p_i) = 0.$$

Hence,

$$(1 - p_i)b_i = (1 - p_i)c_i = b_i(1 - p_i) = c_i(1 - p_i) = 0.$$

Also,

$$y_i(1 - p_i) = (1 - p_i)y_i = 0,$$

since $a_i(1 - p_i) = (1 - p_i)a_i = 0$. We therefore get

$$(*) \qquad\qquad y_i = p_i y_i p_i.$$

Since $\sum a_i = \sum b_i = 1$, we must have $y_1 + y_2 + y_3 = 0$. It follows from $(*)$ that

$$y_1 = \begin{bmatrix} r & r \\ r & r \end{bmatrix}, \; y_2 = \begin{bmatrix} s & \omega s \\ \omega^* s & ss \end{bmatrix} \text{ and } y_3 = \begin{bmatrix} t & \omega^* t \\ \omega t & t \end{bmatrix}.$$

Since $y_1 + y_2 + y_3 = 0$, we get $r + s + t = 0$, $r + \omega^* s + \omega t = 0$ and $r + \omega s + \omega^* t = 0$. This means that

$$r + \omega^* s - \omega(r + s) = 0 \text{ and}$$

$$r + \omega s - \omega^*(r + s) = 0.$$

Hence,

$$r(1 - \omega) = s(\omega - w^*) \text{ and}$$

$$r(1 - \omega^*) = s(\omega^* - \omega),$$

which gives

$$s\frac{\omega - \omega^*}{1 - \omega} = s\frac{\omega^* - \omega}{1 - \omega^*},$$

and therefore either $s = 0$ or $\omega^* - 1 = 1 - \omega$. But in the latter case we would have $2 = \omega^* + \omega = -1$ and so $s = 0$. But then $r = -t$ and $r = -\omega t$; so $r = t = 0$. Hence a is an extreme point, but none of the a_i's is a projection.

The following Theorem generalizes an early result of Dvoretzky, Wald, Wolfowitz [17,Theorem 5]. Indeed, their theorem is equivalent to Theorem 5.3 below in the case where the map Ψ is positive.

THEOREM 5.3. *Suppose $\mathcal{N}$ is a nonatomic abelian von Neumann algebra and Ψ is a weak* continuous map from $\mathcal{N}$ into $\mathbb{C}^n$. If $\{a_j : j = 1, \dots, k\}$ are positive elements in $\mathcal{N}$ such that $a_1 + \cdots + a_k = 1$, then there exist projections $\{p_j : j = 1, \dots, k\}$ in $\mathcal{M}$ such that $p_1 + \cdots + p_k = 1$ and $\Psi(p_j) = \Psi(a_j)$ for each j.*

PROOF. Let $\tilde{\mathcal{N}}$ denote the direct sum of k copies of $\mathcal{N}$ and write $a = a_1 \oplus \cdots \oplus a_k$. Let Q be as defined as above. Let Φ stand for the direct sum of k copies of Ψ and observe that Φ has range in $\mathbb{C}^{kn}$ and is weak* continuous. Since Q is weak* compact and has infinite facial dimension by part ii) of Lemma 5.1, we may apply Corollary 1.6 to get an extreme point p of Q such that $\Phi(p) = \Phi(a)$.

By part i) of Lemma 5.1 p has the form $p = p_1 \oplus \cdots \oplus p_k$, where each p_j is a projection. Moreover $p_1 + \cdots + p_k = 1$ and $\Psi(p_j) = \Psi(a_j)$ for each index j. $\square$

We now turn our attention to the finite atomic case. Recall that we define the constant δ as follows:

$$\delta = \delta(\Psi) = sup\{\|\Psi(q)\|_1 : q \text{ is a minimal projection in } \mathcal{N}\}.$$

To avoid trivialities, we shall assume that $\Psi \neq 0$ so that $\delta > 0$.

CONJECTURE 5.4. *If $\mathcal{N}$ is an atomic abelian von Neumann algebra, then for any positive integer $k > 1$ and any positive elements a_i, $i = 1, \ldots, k$ in $(\mathcal{N}_+)_1$ such that $a_1 + \cdots + a_k = 1$, there exist projections p_i, $i = 1, \ldots, k$ in $\mathcal{N}$ such that $p_1 + \cdots + p_k = 1$ and $\|\Psi(p_i - a_i)\|_\infty < 3\delta$.*

REMARK. It is easy to see that Conjecture 5.4 is false if we replace 3δ by any $\delta' < \delta$. For example, we may take $\mathcal{N} = \mathbb{C}, a_j = 1/k$ for each j and define Ψ by $\Psi(z) = z$. In this case it is clear that our only choice for the p_j's is to take $p_1 = 1$ and $p_j = 0$ for $j \geq 2$. We have $\delta = 1$ and $\Psi(p_1 - a_1) = \frac{k-1}{k}$. Thus, for k large enough, we have $\delta' < \|\Psi(p_1 - a_1)\|_\infty$.

In Theorem 5.7 we present the proof of a special case of this conjecture, namely when all of the a_i are equal to $1/k$ times the identity. We begin with 2 lemmas.

LEMMA 5.5. *If s and k are positive integers such that*

$$2^s < k \leq 2^{s+1},$$

then

(∗)
$$\frac{2}{k} \leq \left(\frac{2}{3}\right)^s.$$

Moreover, if $k \geq 6$, then

(∗∗)
$$\frac{2}{k-1} \leq \left(\frac{2}{3}\right)^s.$$

PROOF. The inequality (∗) is easily verified for pairs (s, k) with the values (1,3), (1,4) and (2,5). Moreover, (∗∗) is true for (2,6), (2,7) and (2,8). So, to finish the proof, it suffices to show that (∗∗) is true for $s \geq 3$ and $k > 8$. Clearly, it is enough to establish (∗∗) when $k - 1 = 2^s$. Now (∗∗) holds in this case if and only if

$$3^s \leq 2^{2s-1} = (1/2)4^s.$$

Since $s \geq 3$, we have

$$3^s = (27)3^{s-3} < (32)4^{s-3} = (1/2)4^s.$$

Hence, (∗∗) is true in this case and the proof is complete. $\square$

The following calculation will be used repeatedly in the proof of the next theorem.

LEMMA 5.6. *If p and q are projections in $\mathcal{N}$ and α, β, and γ are non–negative numbers such that*

  i) $\|\Psi(q - \alpha 1)\|_\infty < \delta$ *and*
  ii) $\|\Psi(p - \beta q)\|_\infty < \gamma\delta$,

then

$$\|\Psi(p - \alpha\beta 1)\|_\infty \leq \delta(\gamma + \beta).$$

PROOF. Multiplying the inequality in i) by β, we get

$$\|\Psi(\beta q - \alpha\beta 1)\|_\infty < \beta\delta.$$

Hence,

$$\|\Psi(p - \alpha\beta 1)\|_\infty < \|\Psi(p - \beta q)\|_\infty + \|\Psi(\beta q - \alpha\beta 1)\|_\infty \leq \gamma\delta + \beta\delta.$$

□

THEOREM 5.7. *Suppose $\mathcal{N}$ is an atomic abelian von Neumann algebra and Ψ is a nonzero weak* continuous map of $\mathcal{N}$ into $\mathbb{C}^n$. If k is a positive integer with $k > 1$, and if s denotes the least positive integer such that $k \leq 2^s$, then there are projections p_i, $i = 1, \ldots, k$ in $\mathcal{N}$ such that $p_1 + \cdots + p_k = 1$ and*

(0) $$\|\Psi(p_i - (1/k)1)\|_\infty < 3\delta(1 - (2/3)^s).$$

In the special case where $k = 2^s$, we get the stronger conclusion

(0′) $$\|\Psi(p_i - (1/k)1)\|_\infty < 2\delta(1 - (1/2)^s).$$

PROOF. Note that since Ψ is nonzero, $\delta > 0$. We proceed by induction on s. Suppose $s = 1$ so that $k = 2$. By part i) of Theorem 4.1 there is a projection q in $\mathcal{N}$ such that

(1) $$\|\Psi(q - (1/2)1)\| < \delta,$$

Clearly, we also have

(1′) $$\|\Psi(1 - q - (1/2)1)\|_\infty < \delta,$$

so that (0′) holds. Assume by induction that the theorem has been proved for all $k \leq 2^s$, where $s \geq 1$. Fix k with $2^s < k \leq 2^{s+1}$. It is convenient to consider 2 cases.

First assume k is even. By using part i) of Theorem 4.1 and arguing as above, we get a projection q in $\mathcal{N}$ such that (1) and (1′) hold. If $q = 0$, then put $p_i = 0$ for $i = 1, \ldots, k/2$. Since $k > 1$ and $q = 0$, we have

$$\|\Psi(p_i - (1/k)1)\|_\infty = \|\Psi((1/k)1)\|_\infty = \frac{2}{k}\|\Psi((q - 1/2)1)\|_\infty < \frac{2\delta}{k},$$

for $i = 1, \ldots, k/2$. Since $2^s < k$ and $s \geq 1$, we have

$$\frac{2}{k} \leq \frac{2}{2^s + 1} < 1.$$

Since $2(1 - (1/2)^{s+1}) > 1$ (resp. $3(1 - (2/3)^{s+1}) > 1$), (0) (resp. (0′)) holds for these projections. If $q \neq 0$, then consider the restriction Ψ_1 of Ψ to $q\mathcal{N}$. We have that $k/2 \leq 2^s$ and $\delta(\Psi_1) \leq \delta(\Psi)$. Hence, by our induction hypothesis, we get projections p_i, $i = 1, \ldots, k/2$ in $(\mathcal{N}_+)_1$ with sum q and such that

(2) $\qquad \|\Psi(p_i - (2/k)q)\|_\infty < 3\delta(1 - (2/3)^s) = \delta \sum_{j=0}^{s-1}(2/3)^j$, if $k < 2^{s+1}$,

or

(2′) $\qquad \|\Psi(p_i - (2/k)q)\|_\infty < 2\delta(1 - (1/2)^s) = \delta \sum_{j=0}^{s-1}(1/2)^j$, if $k = 2^{s+1}$.

Now apply Lemma 5.6 to (1) and (2) with $\alpha = 1/2$, $\beta = 2/k$ and $\gamma = 1 + \cdots + (2/3)^{s-1}$ to conclude that

(3) $\qquad \|\Psi(p_i - (1/k)1)\|_\infty < \delta \left\{ \frac{2}{k} + \sum_{j=0}^{s-1}(2/3)^j \right\}$, if $k < 2^s$.

Similarly, Lemma 5.6 applied to (1′) and (2′) gives

(3′) $\qquad \|\Psi(p_i - (1/k)1)\|_\infty < \delta \left\{ \frac{2}{k} + \sum_{j=0}^{s-1}(1/2)^j \right\}$, if $k = 2^s$.

If $k = 2^{s+1}$, then $2/k = (1/2)^s$ and (0′) follows from (3′) for $i = 1, \ldots, k/2$. Now suppose $k < 2^{s+1}$. Since $2^s < k$, by Lemma 5.5 we have $2/k \leq (2/3)^s$. Hence, (0) follows from (3) for $i = 1, \ldots, k/2$. To show the existence of the remaining $k/2$ projections, we replace q by $1 - q$ and repeat the previous argument. Thus, the Theorem is true if k is even.

Now suppose that k is odd so that $2^s < k < 2^{s+1}$. It is convenient to treat the cases $k = 3$ and $k = 5$ separately. So suppose $k = 3$. By part i) of Theorem 4.1 there is a projection q such that

(4) $\qquad \|\Psi(q - (1/3)1)\|_\infty < \delta < 3\delta(1 - (2/3)^2)$

and

(5) $\qquad \|\Psi((1 - q) - (2/3)1)\|_\infty < \delta.$

Put $p_1 = q$. If $q = 1$, then $\|\Psi((2/3)1)\|_\infty < \delta$ and we may take $p_2 = p_3 = 0$. Otherwise, as above, we may use Theorems 4.1 to get projections p_2, p_3 that sum to $1 - q$ and such that

(6) $\qquad \|\Psi(p_i - (1/2)(1 - q))\|_\infty < \delta$

for $i = 2, 3$. Applying Lemma 5.6 to (5) and (6) with $\alpha = 2/3$, $\beta = 1/2$ and $\gamma = 1$, we get

$$(7) \qquad \|\Psi(p_i - (1/3)1)\|_\infty < \delta(1 + (1/2)) \leq 3\delta(1 - (2/3)^2)$$

for $i = 2, 3$.

Now suppose $k = 5$ and argue as before to get a projection q such that

$$(8) \qquad \|\Psi(q - (2/5)1)\|_\infty < \delta \text{ and}$$

$$(9) \qquad \|\Psi((1 - q) - (3/5)1)\|_\infty < \delta.$$

If $q = 0$, then we have

$$\|\Psi((1/5)1)\|_\infty < \frac{\delta}{2} < 3\delta(1 - (2/3)^2),$$

so that (0) holds with $p_1 = p_2 = 0$. Also, if $1 - q = 0$, then we have

$$\|\Psi((1/5)1)\|_\infty \leq \frac{\delta}{3} < 3\delta(1 - (2/3)^2),$$

so that we may put $p_3 = p_4 = p_5 = 0$ in this case. So, suppose $q \neq 0$ and $1 - q \neq 0$. Arguing as before we may find p_1 and p_2 such that $p_1 + p_2 = q$ and

$$(10) \qquad \|\Psi(p_i - (1/2)q)\|_\infty < \delta.$$

Using the case $k = 3$ for $(1 - q)\mathcal{N}$ and formula (7), we get projections p_i such that

$$(11) \qquad \|\Psi(p_i - (1/3)(1 - q))\|_\infty < (3/2)\delta$$

for $i = 3, 4, 5$. Applying Lemma 5.6 to (8) and (10) we get

$$(12) \qquad \|\Psi(p_i - (1/5)1)\|_\infty < (3/2)\delta$$

for $i = 1, 2$ and, using (9) and (1), we have

$$(13) \qquad \|\Psi(p_i - (1/5)1)\|_\infty < \delta((3/2) + (1/3)) = (11/6)\delta$$

for $i = 3, 4, 5$. Since $(3/2)\delta < (11/6)\delta < 3\delta(1 - (2/3)^3)$, (0) holds for $k = 5$.

Finally, suppose $k \geq 7$. Use Theorem 4.1 to get a projection q such that

$$(14) \qquad \|\Psi(q - ((k - 1)/2k)1)\|_\infty < \delta$$

and

$$(15) \qquad \|\Psi((1 - q) - ((k + 1)/2k)1)\|_\infty < \delta.$$

As above we may assume $q \neq 0$. Apply our induction hypothesis to the restriction of Ψ to $q\mathcal{N}$ to get projections p_i, $i = 1, \ldots, (k-1)/2$, that sum to q and such that

$$(16) \qquad \|\Psi(p_i - (2/(k-1))q)\|_\infty < \delta \sum_{j=0}^{s-1} (2/3)^j.$$

Put $\alpha = (k-1)/2k$, $\beta = 2/(k-1)$, $\gamma = 1 + \cdots + (2/3)^{s-1}$ and apply Lemma 5.6 to (14) and (16) to conclude that

$$(17) \qquad \|\Psi(p_i - (1/(k)q)\|_\infty < \delta \left\{ \frac{2}{k-1} + \sum_{j=0}^{s-1} (2/3)^j \right\}.$$

Since $k > 6$, by $(**)$ of Lemma 5.6 we get $2/(k-1) \leq (2/3)^s$, and (0) holds by (11). To complete the proof we argue as before to get projections p_i, $i = 1 + (k-1)/2, \ldots, k$ such that

$$(18) \qquad \|\Psi(p_i - (1/(k)q)\|_\infty < \delta \left\{ \frac{2}{k+1} + \sum_{j=0}^{s-1} (2/3)^j \right\}.$$

Since $2/(k+1) < 2/(k-1)$, these p_i's also have the desired property. $\square$

REMARKS.

1) If the domain of Ψ is $\mathcal{M}$, where $\mathcal{M}$ is the direct sum of finite type I factors, then we may apply Theorem 5.7 to a MASA $\mathcal{N}$ in $\mathcal{M}$. Hence, the conclusions of Theorem 5.7 hold for $\mathcal{M}$.

2) If $\mathcal{N}$ has finite dimension n and $n < k$, then at least $k - n$ of the p_i's must be 0. On the other hand, if $3\delta \leq 1/k$, then none of the p_i's can be 0.

Let $\mathcal{T}$ denote an infinite binary tree with vertices

$$\{v(s,t) : t = 1, \ldots, 2^s, \ s = 0, 1, \ldots \}.$$

THEOREM 5.8. *If $\mathcal{N}$ is atomic and has infinite dimension, then there exists a set $\{p_v\}$ of projections in $\mathcal{N}$ indexed by the vertices of $\mathcal{T}$ such that the following conditions hold.*

  i) *If v is an ancestor of w in $\mathcal{T}$ then $p_w \leq p_v$.*
  ii) *If v and w are on distinct branches in $\mathcal{T}$ then $p_v p_w = 0$.*
  iii) *For each s,*

$$\sum_{t=1}^{2^s} p_{v(s,t)} = 1.$$

  iv) *If w is a vertex immediately below the vertex v, then*

$$\|\Psi(p_w - (1/2)p_v)\|_\infty \leq \delta.$$

  v) *For each s and t,*

$$\|\Psi(p_{v(s,t)} - (1/2^s)1)\|_\infty < 2\delta(1 - 2^{-s}).$$

PROOF. The family $\{p_v\}$ is constructed by induction on s. If $s = 0$, put $p_{v(0,1)} = 1$. If $s \geq 0$ and $\{p_{v(s,t)} : t = 1, \ldots, 2^s\}$ have been selected satisfying the conditions above, then the $p_{v(s+1,t)}$'s are constructed just as in the proof of Theorem 5.7. In particular, v) is established as in the proof of $(0')$ in Theorem 5.7. $\square$

§6. LYAPUNOV THEOREMS FOR SINGULAR MAPS

Introduction

Recall that a continuous linear functional f on a von Neumann algebra $\mathcal{M}$ is **singular** if and only if for each nonzero projection p in $\mathcal{M}$, there is a nonzero projection $q \leq p$ such that $f(q) = 0$ [1,II.1]. Also, a bounded linear map Ψ from a von Neumann algebra $\mathcal{M}$ into a normed linear space $\mathcal{X}$ is said to be **singular** if $\Psi^*(\mathcal{X}^*)$ consists entirely of singular functionals on $\mathcal{M}$ [42,p.128].

We begin this section by studying singular functionals. We show (Theorems 6.7 and 6.8) that, if the center of the finite part of $\mathcal{M}$ is finite–dimensional (and modest set theoretic assumptions are made so that $\mathcal{M}$ is "essentially countably decomposable"), then we have more precise information about projections in the kernel of a singular state. Specifically, in this case, if f is a singular state on $\mathcal{M}$ and p is a projection in $\mathcal{M}$, there is a projection $q \leq p$ in the kernel of f such that $p - q$ is equivalent to a subprojection of q. Further, if p is properly infinite, then we can also choose q to be equivalent to p. The assumption that the center of the finite part of $\mathcal{M}$ be finite dimensional is necessary for this property to hold, as we show in Theorem 6.9.

We next study a singular map Ψ on $\mathcal{M}$ into a normed linear space $\mathcal{X}$ such that X^* is weak* separable. In Theorem 6.12 we show that a type 1 Lyapunov theorem holds, assuming the center of the finite part of $\mathcal{M}$ has finite dimension (and that $\mathcal{M}$ is "essentially countably decomposable"). Theorem 6.12 may seem counterintuitive on first reading. The range is allowed to be infinite dimensional, whereas for weak* continuous maps, which are usually better–behaved, the range must be finite dimensional for a type 1 Lyapunov conclusion to hold. However, examination of the proofs of our previous Lyapunov theorems shows that the assumption of finite dimensionality on the range of Ψ was used to ensure that this dimension was strictly smaller than the dimension of the domain. Now if f denotes a singular state on $\mathcal{M}$, then the associated representation acts on an inseparable Hilbert space [42,p.352]. It follows that Ψ factors through a singular representation of $\mathcal{M}$, and so the domain of Ψ has an "essentially" inseparable character. For example, if $\mathcal{M} = B(H)$, then Ψ acts on the Calkin algebra, which contains an uncountable family of orthogonal projections. Thus, given the inseparable character of the domain, it is not unreasonable to expect that type 1 Lyapunov theorems hold for singular maps whose ranges satisfy some sort of separability condition.

After proving our Lyapunov theorem for singular maps, we then show that there are many such maps which arise naturally. These include projections of norm one from $B(H)$, for separable H, onto any injective von Neumann subalgebra of $B(H)$ which contains no minimal projections.

Finally, we prove a combination Lyapunov theorem where we assume only that $\Psi : \mathcal{M} \to \mathbb{C}^n$ is a norm continuous linear map. (Actually we can even extend to an affine, norm continuous map on a weak* closed face of $(\mathcal{M}_+)_1$.) Given such a Ψ, we have the decomposition

$$\Psi = \Psi_n + \Psi_s$$

into its normal and singular parts [42,p.128]. Results from earlier sections are applied to Ψ_n while results from this section are applied to Ψ_s. In Theorem 6.23 it is shown that if $\mathcal{M}$ is "essentially countably decomposable" and nonatomic and the center of the finite part of $\mathcal{M}$ is finite dimensional, then $\Psi(P) = \Psi((\mathcal{M}_+)_1)$.

The reader may find it profitable to compare the type 1 results in this section to the type 2 results in Theorem 3.12.

Set Theoretic Preliminaries

In order to derive our stronger characterization of singular functionals, we use techniques from set theory. In the countably decomposible case, the set theoretic arguments that are needed are elementary. It turns out that for "large" von Neumann algebgras, however, much more elaborate machinery is required. In fact, to derive our results, we have found it necessary to impose a technical condition which limits the size of the von Neumann algebra under consideration.[1] In order to define this condition, we begin by reviewing some standard ideas and results from set theory.

It should be emphasized that this material is not needed for the countably decomposable case. Thus, the reader who is not set theoretically inclined may safely skip the next three results and read the remainder of this section under the assumption that all von Neumann algebras under consideration are countably decomposable. Since these next results are purely set theoretic in nature,they will be presented in this more abstract setting. We are grateful to Steve Simpson and Hugh Woodin for several helpful discussions concerning this material .

Recall that an ultrafilter U of subsets of a set S is **nonprincipal** if $\{s\} \notin U$ for each $s \in S$. U is said to be κ-**complete** if it is closed under the formation of intersections of families of fewer than κ elements of U. An uncountable cardinal κ is said to be **measurable** if it admits a κ–complete nonprincipal ultrafilter. By [26,Lemma 27.1] the least cardinal whose power set admits a nontrivial, countably additive two–valued measure is measurable. It is known that if a measurable cardinal exists, then it must be quite large. For example, if we write $\exp_2$ for exponentiation with base 2 and we let c denote the cardinality of the continuum, then

$$\exp_2(\exp_2(\ldots(\exp_2(c))\ldots))$$

---

[1]In practice, most operator algebraists (willing to assume the continuum hypothesis) will never encounter a von Neumann algebra that fails to satisfy the condition. In particular, if $\mathcal{M}$ is countably decomposable, then, assuming the continuum hypothesis, $\mathcal{M}^{**}$ is essentially countably decomposable.

is not measurable for any finite number of compositions.

If S is a set, an **ideal** in the power set of S is a family I of subsets of S such that the following two conditions hold.

    i) I is closed under the formation of finite unions.

    ii) If $\sigma \in I$ and $\tau \subset \sigma$, then $\tau \in I$.

Thus, I is an ideal if and only if $\{S \backslash \sigma : \sigma \in I\}$ is a filter. It is straightforward to check that if μ is a finitely additive measure on the power set of S, then the null sets of μ form an ideal.

An ideal I is said to be σ–**complete** if it is closed under the formation of countable unions. More generally, if κ is an uncountable cardinal, then I is κ–complete if I is closed under the formation of unions of fewer than κ sets. Thus I is σ–complete if and only if I is $\aleph_1$–complete.

An ideal I is said to be σ–**saturated** if it contains every singleton set and if every family of pairwise disjoint sets that are not in I is countable. Suppose μ is a finite, finitely additive measure that is 0 on singleton sets. If $\{\sigma_\alpha\}$ is a pairwise disjoint family of sets, then $\{\alpha : \mu(\sigma_\alpha) > 1/k\}$ is finite for each k and so the ideal of null sets of μ is σ–saturated.

It turns out that the existence of a σ–complete, σ–saturated ideal is intimately related to the existence of a measurable cardinal. In fact we have the following:

    i) If κ is the least cardinal with the property that there is a σ-complete σ–saturated ideal over κ, then the ideal is κ–complete. [26, Lemma 27.3 (b), p. 299]

    ii) If κ carries a κ–complete σ–saturated ideal, then either κ is measurable or $\kappa \leq c$. [26, Corollary, p.301]

    iii) If κ is a successor cardinal, then there is no κ–complete σ–saturated ideal over κ. [26, Lemma 27.8, p. 303].

(If I is an ideal of subsets of a set of cardinality κ, then we say I is an ideal **over** κ or κ **carries** I). The reader is referred to [26] for further details.

Our first result now follows easily. Since we only require that c be a succsessor cardinal, we take the statement "the continuum hypothesis is true" to mean that $c = \aleph_1$, rather than the more general version.

PROPOSITION 6.1. *If κ is less than the first measurable cardinal and the continuum hypothesis is true, then κ does not carry a σ–complete σ–saturated ideal.*

PROOF. Suppose that κ does have a σ–complete σ–saturated ideal. Let λ denote the least cardinal which carries such an ideal. By i), λ carries a λ–complete σ–saturated ideal. Since $\lambda \leq \kappa$, ii) implies that $\lambda \leq c$. But, since $c = \aleph_1$ and λ is uncountable, we have $\lambda = c$ and so λ is a successor cardinal. Thus, iii) gives a contradiction. □

Let us say that a cardinal κ is **submeasurable** if it is less than the first measurable cardinal. Also, we say that **a set S is submeasurable** if its cardinality is submeasurable.

LEMMA 6.2. *Suppose that the continuum hypothesis is true. If S is a submeasurable set and μ is a finitely additive probability measure on the power set* *of S such that $\mu(\{s\}) = 0$ for each s in S, then there exists a sequence $\{\sigma_n\}$ of disjoint subsets of S such that $\mu(\sigma_n) = 0$ for each n and $\cup\sigma_n = S$.*

PROOF. Let us say that a subset σ of S is **good** if $\mu(\sigma) > 0$ and σ admits a countable partition into null sets. We first show that if $\mu(\sigma) > 0$, then σ contains a good subset. Suppose this is not the case, and σ is a subset of S with positive measure that does not contain any good subset. Let I denote the ideal of null sets contained in σ. If $\{\sigma_n\}$ is a sequence of disjoint null sets in σ, then $\mu(\cup\sigma_n) = 0$, since otherwise $\cup\sigma_n$ would be a good subset of σ. Thus, I is σ–complete. Since μ is finite and is 0 on singleton sets, I is σ–saturated. Hence, we get a contradiction by Proposition 6.1, and therefore σ contains a good subset.

Now let $\{\sigma_\alpha\}$ be a maximal family of disjoint good subsets of S. If τ is the complement of $\cup\sigma_\alpha$, then $\mu(\tau) = 0$ since, if not, by the previous paragraph τ would contain a good subset, contradicting the maximality of our family. On the other hand, since $\mu(\sigma_\alpha) > 0$ and μ is a probability measure, there can be at most a countable number of α's. Thus, we can partition each σ_α into a countable number of null sets. These partition sets, together with τ, form a countable partition of S into null sets. $\square$

The techniques used to derive our strengthened property for singular states require that a weak notion of separability hold for the von Neumann algebra. Let us now describe this notion precisely.

DEFINITION 6.3. *A von Neumann algebra $\mathcal{M}$ is **essentially countably decomposable** if, given a singular state f on $\mathcal{M}$, and a family $\{p_\alpha : \alpha \in \kappa\}$ of orthogonal projections in the kernel of f with sum 1, there is a partition of κ into a countable family of disjoint subsets $\{\kappa_n\}$ such that if we write*

$$p_n = \sum_{\alpha \in \kappa_n} p_\alpha,$$

then each p_n lies in the kernel of f.

Note that if $\mathcal{M}$ is any von Neumann algebra and f is a singular state on $\mathcal{M}$, then a simple argument using Zorn's lemma shows that we may always find a family $\{p_\alpha\}$ of orthogonal projections in $\mathcal{M}$ such that $f(p_\alpha) = 0$ for each α and $\sum_\alpha p_\alpha = 1$. If $\mathcal{M}$ is countably decomposable, then the family $\{p_\alpha\}$ must be countable. In general, it may not be the case that we can find such a countable family of p_α's, however. In fact, if a set S admits a nonprincipal ultrafilter $\mathcal{U}$ such that the induced two valued measure on the power set of S is countably additive, then $\mathcal{U}$ induces a singular state on $\ell^\infty(S)$ that does not enjoy the property described in Definition 6.3. Indeed, if $\{\sigma_n\}$ is a sequence of disjoint subsets of S with $\sigma_n \notin \mathcal{U}$ for all n, then by countable additivity, $\cup\sigma_n \notin \mathcal{U}$. In particular $\cup\sigma_n \neq S$ and so the corresponding projections in $\ell^\infty(S)$ cannot sum to the identity. Thus, some restriction on the size of $\mathcal{M}$ may be necessary in order for to ensure that it is essentially countably decomposable.

We now present a result that gives conditions that ensure that $\mathcal{M}$ is essentially countably decomposable. Some more notation is required. Let us say that a von Neumann algebra $\mathcal{M}$ is κ–**decomposable** if every family of orthogonal projections in $\mathcal{M}$ has cardinality less than κ and κ is the least cardinal with this property. Further, we say that $\mathcal{M}$ is **not too large** if it is κ–decomposable for some submeasurable cardinal κ.

PROPOSITION 6.4. *If $\mathcal{M}$ is a von Neumann algebra then $\mathcal{M}$ is essentially countably decomposable if either of the following conditions holds.*

   (1) *$\mathcal{M}$ is countably decomposable.*

   (2) *$\mathcal{M}$ is not too large and the continuum hypothesis is true.*

PROOF. Fix a singular state f on $\mathcal{M}$ and a family $\{p_\alpha : \alpha \in \kappa\}$ of orthogonal projections in the kernel of f which sum to 1. As noted above, If $\mathcal{M}$ is countably decomposable, then κ is countable, and the proof is complete in this case.

Now assume that $\mathcal{M}$ is not too large and that the Continuum Hypothesis holds. We may view κ as a set and define a measure μ on subsets of κ by the formula $\mu(\sigma) = f(p_\sigma)$, where $p_\sigma = \sum_{\alpha \in \sigma} p_\alpha$. It is clear that μ is a finitely additive probability measure and so the required countable decomposition of the identity exists by Lemma 6.2. $\square$

We now turn to some less esoteric set theoretic considerations. If σ and τ are subsets of a set S, then we say σ and τ are **almost disjoint if** $\sigma \cap \tau$ is finite. If π and π' are permutations of the integers $\mathbb{N}$ we say that π and π' are **almost distinct if** $\{n : \pi(n) = \pi'(n)\}$ is finite.

LEMMA 6.5.

   (1) *For any sequence $\{\pi_n\}$ of permutations of $\mathbb{N}$ there exists a permutation π of $\mathbb{N}$ which is almost distinct from each π_n,*

   (2) *There exists an uncountable family of almost distinct permutations of $\mathbb{N}$.*

   (3) *There exists an uncountable family of almost disjoint subsets of N.*

PROOF. Part (1) is proved easily by induction. For part (2) one uses Zorn's Lemma to get a maximal family of almost disjoint permuations. If the family were countable, then by part (1) the family would not be maximal.

To see that (3) is true, use (2) to get an uncountable family Γ of almost distinct permutations. Clearly, it suffices to produce an uncountable family of almost disjoint subsets in any countable set. For each $\pi \in \Gamma$ write

$$\sigma_\pi = \{\, \langle n, \pi(n)\rangle : n \in \mathbb{N}\,\} \subset \mathbb{N} \times \mathbb{N}.$$

It is easy to check that, since the permutations are almost distinct, the sets in $\{\,\sigma_\pi : \pi \in \Gamma\,\}$ are almost disjoint. $\square$

LEMMA 6.6. *Suppose S is a set, μ is a finitely additive probability measure on the power set of S, $\mathcal{I}$ is a countable index set and $\{\sigma_\iota : \iota \in \mathcal{I}\}$ is a (countable) family of pairwise disjoint subsets of S such that $\mu(\sigma_\iota) = 0$ for each $\iota \in \mathcal{I}$. If Γ is an uncountable family of infinite, almost disjoint subsets of $\mathcal{I}$ and for each $\gamma \in \Gamma$ we write*

$$\sigma_\gamma = \cup\{\sigma_\iota : \iota \in \gamma\},$$

then $\mu(\sigma_\gamma) = 0$ for all but a countable number of γ's.

PROOF. For each positive integer k write

$$G(k) = \left\{ \gamma \in \Gamma : \mu(\sigma_\gamma) > \frac{1}{k} \right\}.$$

We claim that each $G(k)$ has cardinality less than k. Indeed, note that, since μ is zero on each σ_ι, if γ and γ' differ by a finite number of elements, then $\mu(\sigma_\gamma) = \mu(\sigma_{\gamma'})$. Thus, if we fix distinct sets $\gamma_1, \ldots, \gamma_n$ in $G(k)$, then, since they are almost disjoint, we may delete a finite number of members from each γ_j to obtain sets $\gamma'_1, \ldots, \gamma'_n$ that are pairwise disjoint and such that $\mu(\sigma_{\gamma_j}) = \mu(\sigma_{\gamma'_j})$. Thus, we have

$$\frac{n}{k} \leq \sum_{j=1}^{n} \mu(\sigma_{\gamma_j}) = \sum_{j=1}^{n} \mu(\sigma_{\gamma'_j}) \leq 1$$

and therefore $n \leq k$. Hence,

$$\{\gamma \in \Gamma : \mu(\tau_\gamma) > 0\} = \cup\{G(k) : k = 2, 3, \ldots\}$$

is countable.   □

Given a family $\{\sigma_\alpha\}$ of sets we say that a set τ is a **transversal** of $\{\sigma_\alpha\}$ if $\tau \cap \sigma_\alpha$ contains exactly one point for each α.

LEMMA 6.7. *If S is a set, μ is a finitely additive probability measure on the power set of S, and $\{\sigma_\alpha : \alpha \in \kappa\}$ a family of disjoint, countable subsets of S such that $\mu(\sigma_\alpha) = 0$ for each α, then there exists a transversal set τ of $\{\sigma_\alpha : \alpha \in \kappa\}$ such that $\mu(\tau) = 0$ if either of the following conditions hold:*

i) *κ is countable.*
ii) *κ is submeasurable and the continuum hypothesis is true.*

PROOF. If κ is countable, we can write $\kappa = \mathbb{N}$, and use Lemma 6.5(2) to get an uncountable family Γ of almost distinct permutations of $\mathbb{N}$. Also, we can enumerate each of the σ_j as $\{s_{jk}\}$ and define a family $\{\tau_\gamma\}_{\gamma \in \Gamma}$ of transversals by $\tau_\gamma = \{s_{j\gamma(j)}\}$. Since the permutations in Γ are almost distinct, the transversals $\{\tau_\gamma\}_{\gamma \in \Gamma}$ are almost disjoint. It follows from Lemma 6.6 that $\mu(\tau_\gamma) = 0$ for some $\gamma \in \Gamma$.

Now suppose ii) holds. By Lemma 6.2 we may find a disjoint sequence $\{\kappa_n\}$ of nonvoid subsets of κ with union κ such that, with $\rho_n = \cup_{\alpha \in \kappa_n} \sigma_\alpha$, we have $\mu(\rho_n) = 0$ for each n. Since each σ_α is countable, for each fixed n we may find a family $\{\tau_{nk}\}$ of transversals for $\{\sigma_\alpha : \alpha \in \kappa_n\}$ such that $\cup\{\tau_{nk} : k \in \mathbb{N}\} = \rho_n$ and the τ_{nk}'s are pairwise disjoint. We are now reduced to finding a permutation π of the natural numbers such that $\mu(\cup\tau_{n\pi(n)}) = 0$. For, if we find such a permutation, then setting $\tau = \cup\tau_{n\pi(n)}$, we get a transversal of the σ_α's which has measure 0. Finding the permutation π is exactly equivalent to proving the Lemma under the assumption that κ is countable. Since this was established in the first paragraph, the proof is complete.   □

Singular Lyapunov Theorems and Related Examples

With these set theoretic preparations completed, we can now return to the study of singular states. Our first result gives the strengthened version of the singularity criterion for states on a properly infinite von Neumann algebra promised on page 50.

THEOREM 6.8. *If $\mathcal{M}$ is a properly infinite, essentially countably decomposable von Neumann algebra and f is a singular state on $\mathcal{M}$, then there is a projection p in $\mathcal{M}$ such that $f(p) = 0$ and $p \sim 1 - p \sim 1$.*

PROOF. We first prove that there is a family of orthogonal properly infinite projections with sum 1 which lie in the kernel of f. It is convenient to treat the purely infinite (i.e., type III) case and the semifinite case separately. So, first suppose $\mathcal{M}$ is a type III von Neumann algebra. By Zorn's Lemma there is a maximal family $\{s_\alpha\}$ of orthogonal properly infinite projections in the kernel of f. If $s = \sum s_\alpha < 1$, then $1 - s$ would dominate a nonzero projection in the kernel of f. Since $\mathcal{M}$ is type III, this projection would automatically be properly infinite. Hence $s = 1$ by the maximality of the family $\{s_\alpha\}$.

Now suppose $\mathcal{M}$ is semifinite. Since $\mathcal{M}$ is essentially countably decomposable, we may find a sequence $\{r_n\}$ of orthogonal projections in the kernel of f which sum to 1. By [42,Theorem 2.15], there is a faithful, normal, semi-finite trace τ on $\mathcal{M}$. Since $\mathcal{M}$ is not finite, we have $\sum \tau(r_n) = \infty$ and so we may form finite sums of the r_n's to get a sequence $\{q_n\}$ of orthogonal projections in the kernel of f with $\tau(q_n) > 1$ for each n. If we now fix an uncountable family Γ of almost disjoint subsets of $\mathbb{N}$ and define $\{q_\gamma : \gamma \in \Gamma\}$ by the formula

$$q_\gamma = \sum_{n \in \gamma} q_n,$$

then each q_γ is properly infinite. By Lemma 6.6 we have $f(q_\gamma) = 0$ for some γ. Hence the kernel of f contains at least one properly infinite projection.

Applying Zorn's lemma once more we get a maximal family $\{s_\beta\}$ of orthogonal properly infinite projections in the kernel of f. Write $s = \sum_\beta s_\beta$. If $r = 1 - s \neq 0$, then r must be finite. Indeed, if not, then r would dominate a properly infinite projection in the kernel of f by the argument of the previous paragraph, and this would again contradict the maximality of the s_β's. As r is finite, $r\mathcal{M}r$ is countably decomposable and so we may write $r = \sum r_n$, where each r_n is a projection in the kernel of f. Since every properly infinite projection is the sum of a sequence of properly infinite projections, we may assume that the family $\{s_\beta\}$ is infinte. Adding the r_n's to distinct s_β's, we get the desired family in this case.

Now suppose $\mathcal{M}$ is properly infinite and essentially countably decomposable. By the argument above we may find a sequence $\{p_n\}$ of properly infinite projections in the kernel of f which sum to 1. For each fixed n, we may use the halving lemma [27,p.412] to find a sequence $\{p_{nk}\}$ of equivalent projections such that $\sum_k p_{nk} = p_n$ and p_{nk} is equivalent to p_n for all k. Define a finitely additive probability measure on the power set of $\mathbb{N} \times \mathbb{N}$ by $\mu(A) = f(\sum_{(i,j) \in A} p_{ij})$ and write $S_n = \{(n,j) : j = 1, \dots \}$. By part i) of Lemma 6.6 there is. a transversal set $T = \{(n,k(n))\}$ for $\{S_n\}$ such that $\mu(T) = 0$. Thus with $p = \sum_n p_{nk(n)}$, we have $f(p) = 0$. Clearly p is equivalent to 1 since p_{nk} is equivalent to p_n for every k and $\sum_n p_n = 1$. We also have that p and $1 - p$ are equivalent, and the proof is complete. $\square$

THEOREM 6.9. *If $\mathcal{M}$ is an essentially countably decomposable von Neumann algebra such that the center of the finite part of $\mathcal{M}$ is finite dimensional and f is a singular state on $\mathcal{M}$, then there is a projection p in $\mathcal{M}$ such that $f(p) = 0$ and $1 - p$ is equivalent to a subprojection of p. Further, if m has no type I_n summand for odd n, then we can choose p to be equivalent to $1 - p$.*

PROOF. By Theorem 6.8 the conclusion holds for the properly infinite part of $\mathcal{M}$. As the center of the finite part of $\mathcal{M}$ has finite dimension, this summand is itself the direct sum of a finite number of finite factors. Clearly we may treat each summand separately. A finite Type I summand presents no problem because it must be finite dimensional and so each singular functional vanishes identically there. Thus, we may assume that $\mathcal{M}$ is a factor of type II_1. In this case $\mathcal{M}$ is countably decomposable, and so we may find a sequence $\{p_n\}$ of orthogonal projections such that $f(p_n) = 0$ for each n and $\sum p_n = 1$. Let τ denote the faithful normal trace on $\mathcal{M}$. As the p_n's sum to 1, there is an integer m such that

$$\tau(p_1 + \cdots + p_m) > 1/2.$$

Since $\mathcal{M}$ is a II_1 factor, $(p_1 + \cdots + p_m)\mathcal{M}(p_1 + \cdots + p_m)$ is nonatomic, so by Theorem 2.5 there is a projection $p < p_1 + \cdots + p_m$ with $\tau(p) = 1/2$. We have

$$0 \le f(p) \le f(p_1 + \cdots + p_m) = 0,$$

and so $f(p) = 0$. Since $\tau(p) = 1/2$, p and $1 - p$ are equivalent. $\square$

We are grateful to M. Takesaki for pointing out a special case of the following theorem which shows the necessity of the assumption that the center of the finite part of $\mathcal{M}$ be small in the last theorem.

PROPOSITION 6.10. *If $\mathcal{M}$ is a finite von Neumann algebra and the center of $\mathcal{M}$ is infinite-dimensional, then $\mathcal{M}$ admits a singular trace f. If q is a projection such that $f(q) > 1/2$, then q is not equivalent to any subprojection of $1 - q$.*

PROOF. Write $\mathcal{Z}$ for the center of $\mathcal{M}$. Since $\mathcal{M}$ is finite, it admits a center valued trace T [42,p.312]. Since $\mathcal{Z}$ is infinite-dimensional, it is not reflexive [42,p.54] so, there is a singular state g on $\mathcal{Z}$ [42,p.127]. (For example, we may may take g to be a weak*limit point of an infinite sequence of orthogonally supported states). If we write $f = g \circ T$, then we claim that f is a singular trace on $\mathcal{M}$. Since f is a composition of traces, it is a trace. To show that f is singular, let p be a nonzero projection in $\mathcal{M}$. Since T is positive and faithful [42, p. 312], $T(p)$ is positive and nonzero. Thus, there is a spectral projection r of $T(p)$ and a positive scalar λ such that $\lambda T(p) \ge r$. Since g is singular, we can choose a nonzero projection s in the center of $\mathcal{M}$ such that $s \le r$ and $g(s) = 0$. Then $T(sp) = sT(p)$ by the module property of T [42,p.312]. Thus,

$$T(sp) = sT(p) \ge \lambda^{-1}sr = \lambda^{-1}s \ne 0,$$

hence $sp \neq 0$. Since g and f are positive and T has norm 1,

$$0 \leq f(sp) = g(T(sp)) = g(sT(p)) \leq g(s) = 0.$$

As s is central, sp is a projection that is nonzero and dominated by p. Thus, f is singular.

Now, if q is a projection in $\mathcal{M}$ that is equivalent to a subprojection of $1 - q$, then there is a partial isometry v in $\mathcal{M}$ such that $v^*v = q$ and $vv^* \leq 1 - q$. Since f is a trace and $f(q) > 1/2$, we get that

$$1/2 \geq f(1) - f(q) = f(1-q) \geq f(vv^*) = f(v^*v) = f(q) > 1/2.$$

This contradiction completes the proof. $\square$

Suppose A is a unital C^*-algebra and $a \in (A_+)_1$. It is well-known (and easy to check) that if we write $b = \sqrt{a - a^2}$ and

$$p = \begin{bmatrix} a & b \\ b & 1-a \end{bmatrix},$$

then p is a projection in the 2×2 matrices over A. Similarly, if $x \in (A)_1$, we set $c = \sqrt{1 - x^*x}$, $d = \sqrt{1 - xx^*}$ and write

$$u = \begin{bmatrix} x & d \\ c & -x^* \end{bmatrix},$$

then u is unitary. The following lemma is an "internal" version of these facts.

LEMMA 6.11. *If q is a projection in a unital C^*-algebra A and if there is a partial isometry w in A such that $w^*w = q$ and $ww^* \leq 1 - q$, then the following statements are true.*

    i) *If $a \in (A_+)_1$, then there is a projection p in A such that $qpq = qaq$*
    ii) *If $b \in (A_{sa})_1$, then there is a symmetry s in A such that $qsq = qbq$.*
    iii) *If $x \in (A)_1$, then there is a unitary u in A such that $quq = qxq$.*

PROOF. Fix a in A with $0 \leq a \leq 1$ and write $c = qaq$, $d = \sqrt{c - c^2}$. Define p by the formula

$$p = c + w(q - c)w^* + wd + dw^*.$$

Clearly, $p = p^*$, and we have

$$\begin{aligned} p^2 &= c^2 + 0 + wcd + 0 + 0 + w((q - c)^2)w^* + 0 + d(q - c)w^* \\ &\quad + 0 + w(q - c)d + 0 + d^2 + cdw^* + 0 + wd^2w^* + 0 \\ &= c^2 + d^2 + w(cd + d - cd) + w((q - c)^2 + d^2)w^* + (d - cd + cd)w^* \\ &= c + wd + w(q - c)w^* + dw^* = p. \end{aligned}$$

Hence p is a projection and, clearly, $qpq = qaq$. So, i) holds.

If b is as in ii) above, write $a = (b+1)/2$ and use the first part of the proof to get a projection p such that $qpq = qaq$. Put $s = 2p-1$. Clearly, s is a symmetry. Since $b = 2a - 1$, we have $qsq = qbq$, as desired.

Finally, if $\| x \| \leq 1$, write

$$c = \sqrt{q - qxqx^*q} \text{ and } d = \sqrt{q - qx^*qxq}$$

Note that if we define the function f by $f(t) = \sqrt{1-t}$, then $c = f(qx^*qxq)$. Moreover, f has a power series expansion of the form

$$f(t) = 1 + \alpha_1 t + \alpha_2 t^2 + \ldots$$

that converges for $|t| \leq 1$. It follows that

$$
\begin{aligned}
cqxq &= (q + \alpha_1 qxqx^*q + \alpha_2(qxqx^*q)^2 \ldots)qxq \\
&= qxq(q + \alpha_1 qx^*qxq + \alpha_2(qx^*qxq)^2 \ldots) \\
&= qxqd.
\end{aligned}
$$

Define u by the formula

$$u = qxq - wx^*w^* + cw^* + wd + 1 - (q + ww^*).$$

A straightforward (but even more lengthy) calculation shows $u^*u = uu^* = 1$, and so u is unitary. Since $quq = qxq$, iii) holds and the proof is complete. $\square$

Now we are ready to state and prove our Lyapunov theorem for singular maps.

THEOREM 6.12 (LYAPUYNOV THEOREM FOR SINGULAR MAPS). *If $\mathcal{M}$ is an essentially countably decomposable von Neumann algebra such that the center of the finite part of $\mathcal{M}$ is finite dimensional, if $\mathcal{X}$ is a normed linear space whose dual space $\mathcal{X}^*$ is weak* separable, and if Ψ is a singular map of $\mathcal{M}$ into $\mathcal{X}$, then the following statements are true.*

   i) *If $a \in (\mathcal{M}_+)_1$, there is a projection p in $\mathcal{M}$ such that $\Psi(p) = \Psi(a)$.*
   ii) *If $b \in (\mathcal{M}_{sa})_1$, there is a symmetry s in $\mathcal{M}$ such that $\Psi(s) = \Psi(b)$.*
   iii) *If $x \in (\mathcal{M})_1$, there is a unitary u in $\mathcal{M}$ such that $\Psi(u) = \Psi(x)$.*
   iv) *If p and q are projections in $\mathcal{M}$ such that the center of the finite part of $q\mathcal{M}q$ is finite dimensional and $x \in (p\mathcal{M}q)_1$, then there is an element w in $E((p\mathcal{M}q)_1)$ with $\Psi(w) = \Psi(x)$.*

PROOF. Fix a weak* dense sequence $\{\phi_m\}$ in $\mathcal{X}^*$ and write $f_m = \phi_m \circ \Psi$. Since Ψ is singular, each f_m is a singular linear functional on $\mathcal{M}$. Also, each f_m is a linear combination of 4 singular states f_{mi} on $\mathcal{M}$. Thus, if y and z are elements of $\mathcal{M}$ such that $f_{mi}(y - z) = 0$ for all m and i, then $\phi_m(\Psi(y - x)) = 0$ for all m, and therefore $\Psi(y) = \Psi(z)$ because the ϕ_m's are weak* dense. Write

$$f = \frac{1}{4} \sum_{m=1}^{\infty} \sum_{i=1}^{4} \frac{1}{2^m} f_{mi}$$

and observe that f is a singular state on $\mathcal{M}$ [42, p127]. By Theorem 6.9, there is a projection q in $\mathcal{M}$ such that $f(q) = 1$ and q is equivalent to a subprojection of $1 - q$. Now fix a, b and x as in i), ii) and iii) above. By Lemma 6.11 we may find p, s and u such that p is a projection, $qpq = qaq$, s is a symmetry, $qsq = qbq$, u is unitary and $quq = qxq$. Since $f(q) = 1$, it follows that $f_m(q) = 1$ for each m. Hence, $f_m(p - a) = f_m(q(p - a)q) = 0$ for all m, and therefore, $\Psi(p) = \Psi(a)$. Similarly, we have $\Psi(s) = \Psi(b)$ and $\Psi(u) = \Psi(x)$.

Finally, to establish that iv) holds, we use part iii) of the Theorem and Lemma 3.11 and argue as in the proof of part iv) of Theorem 3.12. $\square$

We note that, as the proof shows, Theorem 6.12 holds under the slightly weaker hypothesis that $\Psi(\mathcal{M})^*$ is weak* separable or, equivalently, that $\Psi^*(\mathcal{X}^*)$ is weak* separable.

In order to extend this result to faces, we need to define a notion of singularity for norm continuous affine maps defined only on a weak* closed face of $(\mathcal{M}_+)_1$. Suppose $F = [p, q]$ is a weak* closed positive face of the unit ball of the von Neumann algebra $\mathcal{M}$ and Ψ is a norm continuous affine map from F to a Banach space $\mathcal{X}$. We say that Ψ is **singular** if, for every x^* in $\mathcal{X}^*$ and every projection $s \neq p$ in F, there exists a projection $t \neq p$ in F such that $t \leq s$ and $x^*(\Psi(t)) = x^*(\Psi(p))$. Note that if Ψ is linear and $F = [0, 1]$, then this is the usual singularity criterion since $\Psi^*(x^*)(t) = x^*(\Psi(t)) = x^*(\Psi(0)) = \Psi^*(x^*)(0) = 0$ and so $\Psi^*(x^*)$ is singular for every x^* in $\mathcal{X}^*$.

COROLLARY 6.13. *Suppose $\mathcal{M}$ is an essentially countably decomposable von Neumann algebra and $\mathcal{X}$ is a normed linear space with weak* separable dual space. If Ψ is a singular affine map from the weak* closed face $F = [p, q]$ in $(\mathcal{M}_+)_1$ into $\mathcal{X}$, and if the finite part of the algebra $(q - p)\mathcal{M}(q - p)$ has finite dimensional center, then for each $a \in F$ there is a projection t in F such that $\Psi(t) = \Psi(a)$.*

PROOF. First define an affine map from $\mathcal{N} = (q - p)\mathcal{M}(q - p)$ into $\mathcal{M}$ by $\phi(a) = a + p$. Note that ϕ takes $(\mathcal{N}_+)_1$ injectively onto F. Define the map Φ on $(\mathcal{N}_+)_1$ by $\Phi(a) = \Psi(\phi(a)) - \Psi(p)$. This is clearly an affine map which takes 0 to 0, and it is norm continuous. Since the interior of $(\mathcal{N}_+)_1$ in its real linear span is not empty, by Proposition 1.5, we may extend Φ to be a continuous real-linear map on this space. By a standard argument, Φ extends to a continuous complex-linear map on all of $\mathcal{N}$. We claim that the map $\Phi : \mathcal{N} \to \mathcal{X}$ is singular. To see this, fix x^* in $\mathcal{X}^*$. If $s \neq 0$ is a projection in $\mathcal{N}$, then $s + p$ is in F. By the singularity of Ψ as defined above, there is a projection $r \neq p$ in F such that $r \leq s + p$ and $x^*(\Psi(p)) = x^*(\Psi(r))$. Now $\Psi(r) = \Phi(r - p) + \Psi(p)$, so

$$x^*(\Psi(p)) = x^*(\Psi(r)) = x^*(\Phi(r - p) + \Psi(p)) = x^*(\Phi(r - p)) + x^*(\Psi(p))$$

by linearity. Subtracting $x^*(\Psi(p))$, we get

$$0 = x^*(\Phi(r - p)) = \Phi^*(x^*)(r - p).$$

Since $r - p \neq 0$ and $r - p \leq s$, this proves that Φ is singular. By Theorem 6.11 there is a projection s in $\mathcal{N}$ such that $\Phi(a + p) = \Phi(s)$. Thus

$$\Psi(a) = \Phi(a + p) = \Phi(s) = \Psi(s + p).$$

Since $s + p$ is a projection in F, the proof is complete. $\square$

It may now be helpful to look at some examples. In the first we show why abelian von Neumann algebras must be avoided when searching for singular Lyapunov theorems.

EXAMPLE 6.14. If $\mathcal{M}$ is an infinite-dimensional abelian von Neumann algebra, then $\mathcal{M}$ admits a singular pure state f. Since $\mathcal{M}$ is abelian, f is multiplicative and so $f(P) \subset \{0, 1\}$, $f(S) \subset \{\pm 1\}$ and $f(U) \subset \{\lambda : |\lambda| = 1\}$. Thus, in this case, no Lyapunov conclusion of any type is possible.

In Proposition 6.10 it was shown that the conclusion of Theorem 6.9 fails for finite von Neumann algebras with infinite-dimensional centers, and, by the last example, no Lyapunov theorem is possible in the (singular) abelian case. These observations do not rule out the possibility of singular Lyapunov theorems in type II_1 von Neumann algebras. The next example is a very special case.

EXAMPLE 6.15. If $\mathcal{N}$ is a II_1 factor and we put $\mathcal{M} = L^\infty(0, 1) \otimes \mathcal{N}$ and define $f = g \circ T$ as in the proof (and notation) of Proposition 6.9, then $f(P) = [0, 1]$. To see this, let $\tau(\cdot)$ denote the trace on $\mathcal{N}$ and fix $0 < t < 1$. By Theorem 2.5 there is a projection p in $\mathcal{N}$ such that $\tau(p) = t$. Writing $q = 1 \otimes p$, we have $T(q) = 1 \otimes \tau(p)1 = t(1 \otimes 1)$ and so $f(q) = t$.

In fact by [27,8.4.4] more is true: If $\mathcal{N}$ is of type II_1, then the image of $T(P)$ is $(\mathcal{Z}_+)_1$, where $\mathcal{Z}$ denotes the center of $\mathcal{N}$.

It follows from Example 2.7 that there is a weak* continuous map Ψ of $B(H)$ into $\ell^\infty(\mathbb{N})$ (whose dual space is weak* separable) such that $\Psi(P) \neq \Psi((B(H)_+)_1)$. The definition of Ψ was somewhat ad hoc, however. Here is a more natural example (which we record as a theorem) where a type 2 Lyapunov theorem holds, but not a type 1 theorem. See also Example 3.8.

Let H denote a complex separable, infinite dimensional Hilbert space and fix an orthonormal basis for H. Let Ψ be the projection of $B(H)$ onto the operators that are diagonal with respect to this basis. Thus, Ψ is weak* continuous and its range is (isomorphic to) $\ell^\infty(\mathbb{N})$.

THEOREM 6.16. If $\Psi : B(H) \to \ell^\infty(\mathbb{N})$ is the map defined above, then $\Psi(P) \neq \Psi((B(H)_+)_1)$. However, if $a \in (\ell^\infty(\mathbb{N})_+)_1$ and $\epsilon > 0$, then there is a projection $p \in B(H)$ such that $\Psi(p) - a$ is compact and $\|\Psi(p) - a\| < \epsilon$.

PROOF. If $\tau(\cdot)$ denotes the semifinite trace on $B(H)$ then $\tau(P) = \mathbb{N} \cup \{\infty\}$. Hence, if d denotes a positive diagonal operator whose entries sum to $1/2$, then $d \notin \Psi(P)$.

For the latter assertion, first note that by [35,p.220] we have that if

$$\langle \alpha_1, \ldots, \alpha_n \rangle \in (\ell^\infty(n)_+)_1$$

with $\sum_{i=1}^n \alpha_i = m$, where m is an integer, then there is a real $n \times n$ matrix p such that p is a projection and the diagonal entries of p are precisely $\{\alpha_1, \ldots, \alpha_n\}$. Now, if $b = \langle \beta_1, \ldots, \beta_n \rangle \in (\ell^\infty(n)_+)_1$, we may find an element $a = \langle \alpha_1, \ldots, \alpha_n \rangle \in (\ell^\infty(n)_+)_1$ such that $\sum_{i=1}^n \alpha_i$ is an integer and

$\|a - b\|_\infty \leq \frac{1}{2n}$. Thus, there is an $n \times n$ projection p such that

$$\|\Psi_n(p) - b\|_\infty \leq \frac{1}{2n}.$$

(Here Ψ_n denotes the diagonal map of the $n \times n$ matrices).

Returning now to the infinite dimensional case, given $a \in (\ell_+^\infty)_1$ and $\epsilon > 0$, we may write $a = a_1 \oplus a_2 \oplus \cdots \oplus a_n \oplus \cdots$, where each a_n is a finite sequence such that the sum of its entries differ from an integer by less than $\frac{\epsilon}{n}$. Applying the result proved in the previous paragraph to each a_n, we get a sequence $\{p_n\}$ of matrix projections such that the diagonal of p_n differs from a_n by less than $\frac{\epsilon}{n}$. Writing $p = p_1 \oplus p_2 \oplus \cdots \oplus p_n \oplus \cdots$, we have that p is a projection, $\Psi(p) - a$ is compact and $\|\Psi(p) - a\| < \epsilon$. $\square$

We note that, in contrast to Theorem 6.16, we shall show in Corollary 6.19 that if Ψ is a norm 1 projection of $B(H)$ onto a *nonatomic*, abelian subalgebra, then $\Psi(P) = \Psi((B(H)_+)_1)$.

In his original theorem Lyapunov was able to use the weak* continuity of Ψ and the compactness of $(\mathcal{M}_+)_1$ to conclude that $\Psi((\mathcal{M}_+)_1)$ is closed. This need not be the case for singular maps. In fact, Armstrong and Prikery [7,Theorem 1.2,p.500] have shown that $\Psi((\mathcal{M}_+)_1)$ need not be closed even when the dimension of the range of Ψ is 2.

COROLLARY 6.17. *Suppose $\mathcal{M}$ is a von Neumann algebra for which the hypotheses of Theorem 6.12 are satisfied and π is a singular representation of $\mathcal{M}$ on a Hilbert space H.*

(1) *If p is a projection in $B(H)$ such that pH is separable and Ψ is defined by $\Psi(x) = p\pi(x)p$, then i), ii), iii) and iv) of Theorem 6.12 hold for Ψ.*

(2) *If π is irreducible, q is a projection of finite rank in $B(H)$, $a \in B(H)_{sa}$ and $\|a\| < 1$, then there is a symmetry $s \in \mathcal{M}$ such that $aq = \pi(s)q$*

PROOF. The map Ψ is singular because π is singular. Moreover, $\Psi(\mathcal{M}) \subset B(pH)$ and $B(pH)^*$ is separable in its weak* topology because pH is separable. Hence, the hypotheses of Theorem 6.12 are satisfied, and so its conclusions hold for Ψ. Thus, (1) is true.

For (2) note that by [37,2.74] there is an element $b \in \mathcal{M}_{sa}$ such that $\|b\| < 1$ and $\pi(b)q = aq$. Since q has finite rank, there is a projection p of countable rank such that $pq = q$ and $p\pi(b)pq = \pi(b)q$. The existence of the required symmetry now follows from part (1). $\square$

It follows from the next result that there are many other interesting singular maps on $B(H)$.

PROPOSITION 6.18. *Suppose that A is a C^*-subalgebra of a C^*–algebra M, that A has no finite dimensional hereditary C^*-subalgebras, and that $\pi : M \to A$ is a projection of norm 1. If p is a projection in M such that pMp is 1–dimensional, then $\pi(p) = 0$.*

PROOF. Suppose that p is a projection in M such that pMp is 1–dimensional, but $\pi(p) = a \neq 0$. Since π is a projection of norm 1, by [42,p.131] we have

$a \in (A_+)_1$ and $\pi(pb) = ab$ for every b in A. Write q for the spectral projection of a (computed in A^{**}, cf. [37,3.8]) corresponding to the interval $(\frac{1}{2}\|a\|, \|a\|]$, and let C denote the hereditary C^*–algebra $\{b \in A : qbq = b\}$ of A [37,3.11.10]. Define a function θ on $\mathbb{R}$ by the formula

$$\theta(t) = \begin{cases} 1 \text{ if } t \geq \frac{1}{4} \| a \| \\ 0 \text{ if } t \leq \frac{1}{5} \| a \| \\ \text{linear elsewhere} \end{cases}$$

and put $\rho(t) = \theta(t)/t$. If we write $b = \theta(a)$ and $c = \rho(a)$, then we have $ac = b$. Also, $bq = q$, so that b is a relative unit for C. Thus,

$$\pi(pcC) = acC = bC = C.$$

Let $\mathcal{X}$ denote the norm closure of pcC. Since pMp has dimension 1 and M is a C^*–algebra, pM is a minimal right ideal in M; hence, it is isometric to a (possibly finite dimensional) Hilbert space [2,Theorem III.1]. Thus, since $\mathcal{X}$ is a closed subspace of pM, $\mathcal{X}$ is reflexive. Since $\pi(pcC) = C$, C is norm closed and $\mathcal{X}$ is the norm closure of pcC, π takes $\mathcal{X}$ onto C. Therefore we may apply the open mapping theorem to the restriction of π to $\mathcal{X}$ to get a constant $\mu > 0$ such the closed unit ball of C lies in the image of the closed μ-ball of $\mathcal{X}$. Since $\mathcal{X}$ is reflexive, its μ-ball is weakly compact and therefore, since π is weakly continuous, the unit ball of C is weakly compact. But this is impossible because C is an infinite dimensional C^*–algebra [42,p.54]. Hence $\pi(p) = 0$. □

COROLLARY 6.19. *If $\mathcal{R}$ is an injective, nonatomic von Neumann algebra acting on a separable Hilbert space H, and if Ψ is a conditional expectation of $B(H)$ onto $\mathcal{R}$, then the conclusions* i), ii), iii) *and* iv) *of Theorem 6.12 hold for Ψ.*

PROOF. Since $\mathcal{R}$ is nonatomic, it does not contain an hereditary C^*-subalgebra of finite dimension. Hence, by Proposition 6.18 the kernel of Ψ contains the compact operators, so Ψ is singular. Since H is separable, the dual space of $\mathcal{R}$ is weak* separable. Thus the hypotheses of Theorem 6.12 are satisfied for Ψ; so, the conclusions of Theorem 6.12 hold for Ψ. □

We next consider the case where the map Ψ is norm continuous, but not necessarily singular and the range has finite dimension. In this case, we may decompose Ψ as $\Psi_n + \Psi_s$, where Ψ_n is normal (hence, weak* continuous) and Ψ_s is singular [42,p.128]. If the domain $\mathcal{M}$ satisfies the hypotheses of both Theorems 2.5 and 6.12, a Lyapunov theorem of type 1 holds for each of Ψ_n and Ψ_s. Thus, it is natural to ask if a type 1 result holds for Ψ. It turns out that this is indeed the case. Before presenting this theorem, it is useful to review some ideas from convexity theory which will be used in the proof.

Suppose Q is a non–void convex subset in $\mathbb{C}^n$. Clearly, the linear span of $Q - v, v \in Q$ does not depend on the particular choice of v. The **affine hull** of Q is by definition $v + \text{span}(Q - v)$ for any $v \in Q$. The **dimension** of Q is defined to be the dimension of $\text{span}(Q - v)$. Also, the **relative interior** of Q is defined to be the interior of Q when it is regarded as a subset of its affine hull. It is denoted by ri(Q). In particular, if Q contains 0 and spans $\mathbb{R}^n$, then the relative interior of Q is an open set in $\mathbb{R}^n$. By [38,Theorem 6.4] we have that $v \in \text{ri}(Q)$ if and only if for each y in Q there is $\alpha > 0$ such that $(1+\alpha)v - \alpha y \in Q$. Thus, in the notation of Section 1 we have

$$\text{ri}(Q) = \{v \in Q : G(Q,v) = Q\}.$$

We shall use the following result [38,Theorem 6.3]:

$$\text{ri}(Q)^- = Q^- \text{ and } \text{ri}(Q^-) = \text{ri}(Q).$$

First we prove an easy lemma about singular maps and separability. It will be used at two key points in the proof of Theorem 6.23.

LEMMA 6.20. *If $\mathfrak{X}$ is a Banach space, $\Psi : \mathcal{M} \to \mathfrak{X}$ is a singular map such that $\Psi^*(\mathfrak{X}^*)$ is weak* separable, and p is a projection in $\mathcal{M}$, then there is an increasing net $\{p_\alpha\}$ of projections with supremum p such that, for each α, the restriction of Ψ to the algebra $p_\alpha \mathcal{M} p_\alpha$ is identically 0.*

PROOF. As in the proof of Theorem 6.12 we may define a singular functional f such that if $a \geq 0$ and $f(a) = 0$, then $\Psi(a) = 0$. As f is singular, we may use Zorn's lemma to find an increasing net $\{p_\alpha\}$ of projections such that $p_\alpha \to p$ and $f(p_\alpha) = 0$ for all α. For each fixed α, we have that the kernel of f contains $\{a : 0 \leq a \leq p_\alpha\}$ and so Ψ is also 0 on this set. Since this set spans $p_\alpha \mathcal{M} p_\alpha$, Ψ has the desired property. □

Now we prove our "combination theorem". In order to use the result of Theorem 2.5 we must assume that $\Psi : \mathcal{M} \to \mathbb{C}^n$ and that $\mathcal{M}$ is nonatomic. In order to use the result of Theorem 6.12, we must assume that $\mathcal{M}$ is essentially countably decomposable and that we can avoid dealing with a infinite dimensional center in the finite part of a von Neumann algebra. For countably decomposable algebras, Theorem 6.23 applies to all properly infinite nonatomic von Neumann algebras and type II_1 factors. Theorem 6.23 does not apply to abelian algebras. Note that, by Theorem 3.12, a type 2 theorem, analogous to Theorem 6.23, holds for type I_∞, type II and type III von Neumann algebras.

It is convenient to isolate parts of the proof as separate propositions.

PROPOSITION 6.21. *If Ψ is a norm continuous linear map of a von Neumann algebra $\mathcal{M}$ into $\mathbb{C}^n$, then the following statements hold.*

(a) $\Psi = \Psi_n + \Psi_s$, where Ψ_n is normal and Ψ_s is singular.

If, in addition, $\mathcal{M}$ is an essentially decomposable, nonatomic von Neumann algebra such that the center of the finite part of $\mathcal{M}$ has finite dimension, then

(b) *If $a \in (\mathcal{M}_+)_1$ and $\Psi_n(a) \in \text{ri}(\Psi_n((\mathcal{M}_+)_1))$, then there is a projection $p \in \mathcal{M}$ such that $\Psi(p) = \Psi(a)$.*

PROOF. We may write $\Psi(x) = (f_1(x), \ldots, f_n(x))$, where each f_j is a continuous linear functional on $\mathcal{M}$. Further we have the decomposition $f_j = g_j + h_j$ into its normal and singular parts [42, p.127]. Thus, if we write $\Psi_n(x) =$

$(g_1(x), \ldots, g_n(x))$ and $\Psi_s(x) = (h_1(x), \ldots, h_n(x))$, then Ψ_n is normal, Ψ_s is singular and

$$(0) \qquad \qquad \Psi = \Psi_n + \Psi_s.$$

Thus, (a) is true.

For (b), fix $a \in (\mathcal{M}_+)_1$ such that $\Psi_n(a) \in \mathrm{ri}(\Psi_n((\mathcal{M}_+)_1))$. We seek a projection p such that $\Psi(p) = \Psi(a)$. The proof of the existence of such a p is somewhat lengthy. Let us begin by applying Theorem 6.12 to get a projection q with $\Psi_s(q) = \Psi_s(1-a)$. By Lemma 6.20, we may find an increasing net $\{q_\alpha\}$ of projections in the kernel of Ψ_s such that $\lim q_\alpha = 1 - q$. Thus, putting $p_\alpha = q + q_\alpha$, we get an increasing net of projections that converges to 1 and such that

$$(1) \qquad \qquad \Psi_s(1 - p_\alpha) = \Psi_s(a)$$

for each α. Write

$$C = \bigcup_\alpha \mathrm{ri}(\Psi_n(p_\alpha((\mathcal{M}_+)_1)p_\alpha)).$$

We have that C is a convex set whose closure is $\Psi_n((\mathcal{M}_+)_1)$. Since the convex sets $\mathrm{ri}(\Psi_n(p_\alpha((\mathcal{M}_+)_1)p_\alpha))$ are increasing with α and their union is dense in $\mathrm{ri}(C)$, it follows by the results mentioned prior to Lemma 6.20 that

$$\dim(\mathrm{ri}(\Psi_n(p_\alpha((\mathcal{M}_+)_1)p_\alpha)))=\dim(\mathrm{ri}(C))$$

for large α. This means that C is the union of open sets; i.e., $\mathrm{ri}(C) = C$.

By the results mentioned prior to Lemma 6.20, $\Psi_n((\mathcal{M}_+)_1)$ and C have the same relative interior, which is C itself by the previous paragraph. Thus, $\Psi_n(a) \in C$ and so $\Psi_n(a) \in \mathrm{ri}(\Psi_n(p_\alpha((\mathcal{M}_+)_1)p_\alpha))$ for all large α. Since $\lim \Psi_n(1-p_\alpha) = 0$, Ψ_n is weak* continuous and C is open (in the affine hull of C), we have

$$\Psi_n(a) - \Psi_n(1 - p_\alpha) \in \mathrm{ri}(\Psi_n((\mathcal{M}_+)_1)) = C$$

for all large α. Hence, we may fix α such that

$$\Psi_n(a) - \Psi_n(1 - p_\alpha) \in \mathrm{ri}(\Psi_n(p_\alpha((\mathcal{M}_+)_1)p_\alpha)).$$

By Lemma 6.20 there is an increasing net $\{r_\beta\}$ of projections with $\lim r_\beta = p_\alpha$ such that the restriction of Ψ_s to $r_\beta \mathcal{M} r_\beta = 0$ for all β. Write

$$D = \bigcup_\beta \mathrm{ri}(\Psi_n(r_\beta((\mathcal{M}_+)_1)r_\beta)).$$

As with C above, D is an open set in $\Psi(\mathcal{M})$ and $\mathrm{ri}(\Psi_n(r_\beta((\mathcal{M}_+)_1)r_\beta))$ is open in $\Psi(\mathcal{M})$ for each large β. Further, the closure of D is clearly the same as the closure of $\Psi_n(p_\alpha((\mathcal{M}_+)_1)p_\alpha)$ since $\lim_\beta r_\beta = p_\alpha$. Hence these two sets have the same relative interior, namely D itself. Since

$$\Psi_n(a) - \Psi_n(1 - p_\alpha) \in \mathrm{ri}(\Psi_n(p_\alpha((\mathcal{M}_+)_1)p_\alpha)) = D,$$

there exists a fixed β such that

$$\Psi_n(a) - \Psi_n(1 - p_\alpha) \in \mathrm{ri}(\Psi_n(r_\beta((\mathcal{M}_+)_1)r_\beta)).$$

Next use Theorem 2.5 to choose a projection $t \leq r_\beta$ such that

(2) $$\Psi_n(t) = \Psi_n(a) - \Psi_n(1 - p_\alpha).$$

By the choice of the r_β,

(3) $$\Psi_s(t) = 0.$$

If we put

(4) $$p = t + 1 - p_\alpha,$$

then p is a projection, and we get

$$\begin{aligned}
\Psi(p) &= \Psi_n(p) + \Psi_s(p) && \text{by (0)} \\
&= \Psi_n(t + 1 - p_\alpha) + \Psi_s(t + 1 - p_\alpha) && \text{by (4)} \\
&= \Psi_n(a) - \Psi_n(1 - p_\alpha) \\
&\quad + \Psi_n(1 - p_\alpha) + \Psi_s(t) + \Psi_s(1 - p_\alpha) && \text{by (2)} \\
&= \Psi_n(a) + \Psi_s(a) = \Psi(a) && \text{by (3),}
\end{aligned}$$

as desired.   □

Recall from Theorem 2.1 that a weak* closed face of $(\mathcal{M}_+)_1$ has the form $[p, q]$ for projections p and q in $\mathcal{M}$. The method of proof we use in Theorem 6.23 below requires us to state the Theorem for norm continuous affine maps defined on such faces. The content of the next Proposition is that we may always reduce from the case where Ψ is an affine map on a face to the case where Ψ is a linear map on a von Neumann agebra.

PROPOSITION 6.22. *If $F = [p, q]$ is a weak* closed face in the positive unit ball of a von Neumann algebra $\mathcal{M}$ and Ψ is a norm continuous affine map of F into $\mathbb{C}^n$, then the following statements are true.*

   (1) *If we define $\varphi : ((q - p)\mathcal{M}(q - p)_+)_1 \to F$ by $\varphi(a) = a + p$, then φ is an affine order isomorphism of $((q - p)\mathcal{M}(q - p)_+)_1$ onto F .*
   (2) *There is a norm continuous linear map $\Phi : \mathcal{M} \to \mathbb{C}^n$ such that $\Phi = \Psi \circ \varphi^{-1}$ on $((q - p)\mathcal{M}(q - p)_+)_1$.*
   (3) *The map Φ decomposes as $\Phi = \Phi_n + \Phi_s$ into its normal and singular parts and, pulling back by φ, we have $\Psi = \Psi_n + \Psi_s$, where Ψ_n (resp. Ψ_s)) is a normal (resp. singular) map of the face F.*

PROOF. Statement (1) is just part (1) of Proposition 2.2 and (2) is proved exactly as in the proof of Corollary 6.13. We have the decomposition $\Phi = \Phi_n + \Phi_s$ by Proposition 6.21. The map $\Psi_n = \Phi_n \circ \varphi$ is normal because Φ_n is normal and φ is an order isomorphism. Finally, $\Psi_s = \Phi_s \circ \varphi$ is shown to be singular (as a face map) as in the proof of Corollary 6.13.   □

With these preparations complete we may now present the theorem promised above.

THEOREM 6.23. *If $\mathcal{M}$ is an essentially countably decomposable, nonatomic von Neumann algebra, and if Ψ is a norm continuous affine map from a weak* closed face $F = [p, q]$ of $(\mathcal{M}_+)_1$ to $\mathbb{C}^n$ such that the center of the finite part of $r\mathcal{M}r$ is finite dimensional for each nonzero projection $r \leq (q - p)$, then for each a in F there is a projection p in F such that $\Psi(p) = \Psi(a)$.*

PROOF. By part (3) of Proposition 6.22 we may decompose Ψ as the sum of its normal and singular parts, Ψ_n and Ψ_s. The proof is by induction on the dimension of $\Psi_n((F))$. If this dimension is zero, i.e. if $\Psi_n((F)))$ consists of a single point, then Ψ is singular. Indeed, if $s \neq p$ is a projection in F and x^* is a linear functional on $\mathbb{C}^n$, then there is a projection $t \neq p$ with $t \leq s$ and $x^*(\Psi_s(t)) = x^*(\Psi_s(p))$. Since Ψ_n is constant by assumption, we have $\Psi_n(t) = \Psi_n(p)$ and so Ψ is singular. Thus the Theorem follows from Theorem 6.13 in this case. Now assume that the Theorem is true whenever the dimension of $\Psi_n((F)_1)$ is less than n, and suppose that the dimension of $\Psi_n(F))$ is n. Applying Proposition 6.22, it follows that it suffices to prove the theorem in the case where Ψ is linear and $F = (\mathcal{M}_+)_1$.

By part (b) of Proposition 6.21 we have that if $a \in (\mathcal{M}_+)_1$ and $\Psi_n(a) \in$ ri$(\Psi_n((\mathcal{M}_+)_1))$ then there is a projection s such that $\Psi(s) = \Psi(a)$.

To complete the proof fix $b \in (\mathcal{M}_+)_1$ such that

$$(*) \qquad\qquad \Psi_n(b) \notin ri(\Psi_n((\mathcal{M}_+)_1)).$$

In this case if we write $G = G(\Psi_n((\mathcal{M}_+)_1), \Psi_n(b))$, then G is a proper face of $\Psi_n((\mathcal{M}_+)_1)$ by $(*)$ above and the remark prior to Lemma 6.20. Moreover, $\Psi_n(b)$ is in the relative interior of this face by Theorem 1.2. By [38,Theorem 6.3] we have that G and its closure have the same relative interior. Thus the closure G^- is also a proper face. Since G^- is proper, it must have dimension strictly less than n by our induction hypothesis. If we put $H = \Psi_n^{-1}(G^-)$, then H is a weak* closed face of $(\mathcal{M}_+)_1$. Since the restriction of Ψ to H satisfies the hypotheses of the theorem, and since the dimension of $\Psi(H) = G^-$ is less than n, our induction hypothesis allows us to conclude that there exists a projection $s \in H$ such that $\Psi(s) = \Psi(b)$, the proof is complete. $\square$

In [15,Theorem 2] Choda, et al. prove a version of the next Theorem which is slightly stronger in that it applies to arbitrary type II_1 algebras, but which is also weaker since it requires that the state g appearing in the Theorem must be normal. The proof of Theorem 6.24 exploits the geometric structure of convex sets in $\mathbb{R}^2$ and can be generalized to any finite family of states.

THEOREM 6.24. *Suppose $\mathcal{M}$ is an essentially countably decomposable, nonatomic von Neumann algebra such that the center of the finite part of $\mathcal{M}$ is finite dimensional and $\alpha \in (0, 1)$. If f and g are states of $\mathcal{M}$ such that $f(p) = \alpha$ whenever $g(p) = \alpha$ and p is a projection, then $f = g$.*

PROOF. Define $\Psi : \mathcal{M} \to \mathbb{C}^2$ by $\Psi(a) = (g(a), f(a))$. Clearly we have that $\Psi((\mathcal{M}_+)_1)$ is a convex subset of the unit square in the first quadrant of $\mathbb{R}^2$ that contains the points $\Psi(0) = (0, 0)$ and $\Psi(1) = (1, 1)$. If $\Psi((\mathcal{M}_+)_1)$ consists only of the line segment joining these two points, then clearly $f = g$ on all of $(\mathcal{M}_+)_1$, hence on all of $\mathcal{M}$. If not, then the vertical line in $\mathbb{R}^2$ with Cartesian equation $x = \alpha$ must meet $\Psi((\mathcal{M}_+)_1)$ at some point (α, β), where $\beta \neq \alpha$. (This is clear from the picture of $\Psi((\mathcal{M}_+)_1)$). By Theorem 6.23, $\Psi(P) = \Psi((\mathcal{M}_+)_1)$, so there is a projection p such that $\Psi(p) = (g(p), f(p)) = (\alpha, \beta)$. This contradicts the assumption that $f(p) = \alpha$ whenever $g(p) = \alpha$. $\square$

Theorem 6.23 avoids finite von Neumann algebras with infinite dimensional centers. A reasonable alternative approach, which is not developed in the present paper, would be to place more restrictions on the map Ψ instead of on the algebra $\mathcal{M}$. Even if $\mathcal{M}$ is abelian, this approach can lead to a Lyapunov type theorem. For example, it follows from the work of Armstrong and Prikery [7,Theorem 2.2] that the following is true.

> If $\mathcal{M}$ is an abelian von Neumann algebra, $f_1, \ldots, f_n$ are in $\mathcal{M}^*$ such that $|f_j|$ does not dominate any multiple of any pure state (i.e. each f_j is diffuse), and if $\Psi : \mathcal{M} \to \mathbb{C}^n$ is given by $\Psi(a) = (f_1(a), \ldots, f_n(a))$, then $\Psi(P)$ is a convex set.

As noted above, it is also shown in [7,Theorem 1.2] that if $\mathcal{M}$ is infinite dimensional, then $\Psi(P)$ need not be compact. In view of this, the following conjecture seems reasonable.

CONJECTURE 6.25. *If $\mathcal{M}$ is a von Neumann algebra, $f_1, \ldots, f_n$ are diffuse functionals in $\mathcal{M}^*$, and if $\Psi : \mathcal{M} \to \mathbb{C}^n$ is given by $\Psi(a) = (f_1(a), \ldots, f_n(a))$, then $\Psi(P) = \Psi((\mathcal{M}_+)_1)$.*

§7. NONCOMMUTATIVE RANGE

Let H denote a complex, separable Hilbert space of finite or infinite dimension and fix an atomic MASA $\mathcal{D}$ and a projection p in $B(H)$. Throughout this section we shall study the map $\Phi : \mathcal{D} \to B(H)$ defined by the formula

$$\Phi(x) = pxp.$$

We believe that a theorem analogous to part ii) of Theorem 4.1 is true for Φ. Note that in contrast with the setting of Theorem 4.1, the range of Φ is no longer commutative. Thus, it is natural to replace the ℓ^1 and ℓ^∞ norms used in 4.1 with the trace and operator norms and to define δ by the formula

$$\delta = \delta(p) = \sup\{\operatorname{tr}(\Phi(q)) : q \text{ is a minimal projection in } \mathcal{D}\}.$$

(Here $\operatorname{tr}(\cdot)$ stands for the trace on $B(H)$. In the definition of δ, $\Phi(q)$ has rank one and is positive; thus $\operatorname{tr}(\Phi(q)) = \|\Phi(q)\|$.) We believe that the following analog of Theorem 4.1 is true.

CONJECTURE 7.1. *For each self–adjoint element b in $(\mathcal{D})_1$ there exists a symmetry s in $\mathcal{D}$ such that $\|\Phi(s - b)\| \leq 2\delta$.*

If conjecture 7.1 is true, then a problem raised by Kadison and Singer [28] has a positive answer: Each pure state on $\mathcal{D}$ has a unique (pure) state extension to $B(H)$. The conjecture above may be strictly stronger than a positive answer to the Kadison–Singer problem. In fact, each of the following conjectures suffice to yield the result.

CONJECTURE 7.1.1. *There exists a symmetry s in $\mathcal{D}$ such that*

$$\|\Phi(s)\| \leq 2\delta.$$

(I.e., Conjecture 7.1 holds when $b = 0$.)

CONJECTURE 7.1.2. *There exists a constant γ with $0 < \gamma \leq 1/2$ such that, if $\delta(p) < \gamma$, then there exists a symmetry s in $\mathcal{D}$ such that $\|\Phi(s)\| < 1$.*

CONJECTURE 7.1.3. *There are constants γ and ϵ with $0 < \gamma \leq 1/2$ and $0 < \epsilon < 1/2$ such that, if H is finite dimensional and $\delta(p) < \gamma$, then $\|\Phi(s)\| < 1 - 2\epsilon$ for some symmetry s in $\mathcal{D}$.*

REMARK. We shall see in Proposition 7.6 below that the truth of Conjecture 7.1.3 implies the truth of Conjecture 7.1.2. The converse also holds. In fact more is true: If Conjecture 7.1.2 is true for all projections p whose diagonal is compact, then Conjecture 7.1.3 is true.

To see this, suppose that 7.1.3 is false and $\gamma \leq \frac{1}{2}$ is given. In this case, for each positive integer n, we may find a finite rank projection p_n with $\delta(p_n) < \frac{\gamma}{n+1}$ and such that $\|\Psi(s)\| \geq 1 - \frac{1}{n}$ for every diagonal symmetry s. If we write p for the direct sum of the p_n's, then the diagonal of p is clearly compact, $\delta(p) < \gamma$, and $\|\Psi(s)\| = 1$ for every diagonal symmmetry s.

Thus, in view of Proposition 7.6, if Conjecture 7.1.2 can be verified for projections with compact diagonals, then the conjecture holds for arbitrary projections.

In order to clarify the relationship between these conjectures and the problem of Kadison and Singer, it is useful to present some facts about matrices. We use $M_n(\mathbb{C})$ to denote the complex $n \times n$ matrices.

PROPOSITION 7.2. *Suppose that p is a projection in $M_n(\mathbb{C})$ of rank r and p has the form*

$$p = \begin{bmatrix} b & x^* \\ x & c \end{bmatrix},$$

where b is a $k \times k$ matrix, which also has rank r, and $\|b\| < 1$. In this case there is an $(n - k) \times k$ matrix w such that, if we write

$$e = w^*w \quad \text{and} \quad f = ww^*,$$

then e and f are the range projections of b and c respectively. In particular the rank of c is also r. Moreover, we have

$$b = e - w^*cw \quad \text{and} \quad \|c\| = \|e - b\| < 1.$$

PROOF. Since p is a projection, we have

(1) $$x^*x = b - b^2,$$

(2) $$xb + cx = x, \quad \text{and}$$

(3) $$xx^* = c - c^2.$$

Put $d = |x|$. Using (1) we get $d = \sqrt{b - b^2}$. Since $\|b\| < 1$, b and d have the same range and therefore the same rank. Write d' for the inverse of the $k \times k$ matrix d on the complement of its null space. Thus, we have $dd' = d'd = e$, where e is the projection onto the range of b, which is also the range of d. Next put $w = xd'$, and observe that, since x is an $(n - k) \times k$ matrix and d' is a $k \times k$ matrix, this product is well-defined and w is an $(n - k) \times k$ matrix. We have

$$w^*w = d'x^*xd' = d'd^2d' = e.$$

It follows (by adding columns or rows whose entries are all 0 to make w into a square matrix, for example) that $f = ww^*$ is also a projection of rank r, and therefore $n - k \geq r$. Also, since $x = xe$ by (1), $x = wd$. Hence, we may rewrite (2) as

$$wdb + cwd = wd$$

and multiply by d' to get

$$wb = (1 - c)w,$$

where we have used the fact that $db = bd$. Next, multiplying by w^* gives

$$(4) \qquad b = e - w^* cw,$$

as desired. Finally we may rewrite equation (3) above as

$$(5) \qquad wd^2 w^* = c - c^2,$$

from which it follows that $c - c^2$ has rank r. Since c is a submatrix of p, its rank is at most r. Thus

$$r = \operatorname{rank}(c - c^2) \leq \operatorname{rank}(c) \leq r,$$

and so $fc = c$ by (5). Since the rank of f is r, f is the range projection of c. We thus have $\|c\| = \|w^* cw\|$ and therefore by (4), $\|c\| = \|e - b\| < 1$. □

Remarks.

1) In Proposition 7.2, if b has norm 1, then one can reduce to the given situation by restricting to the complement of the null space of $1 - b$.

2) In the notation of Proposition 7.2 suppose that $n - k \geq k$. (Note that we can always arrange this, permuting rows and columns to interchange b with c if necessary.) We may then find an $(n - k) \times (n - 2k)$ matrix w' such that the $n \times n$ matrix $u = [w, w']$ is unitary. We have then that

$$u^* cu = w^* cw + w^* cw' + w'^* cw + w'^* cw'$$
$$= e - b + w^* cw' + w'^* cw + w'^* cw'.$$

The latter 3 terms are 0. For example $w^* cw' = w^* cfw' = w^* cww^* w'$ and $w^* w' = 0$ because u is unitary. Thus, $u^* cu = e - b$, and if we conjugate p by

$$\begin{bmatrix} 1 & 0 \\ 0 & u \end{bmatrix},$$

we get a projection of the form

$$\begin{bmatrix} b & d \\ d & f - b \end{bmatrix}.$$

LEMMA 7.3. *If p and q are projections in a C^*-algebra A, then*

$$\|p(2q - 1)p\| = \max\{2\|pqp\| - 1, \ 1 - 2\|p(1 - q)p\|\}.$$

Moreover, if $0 < \epsilon < 1/2$, then

$$\|p(2q - 1)p\| < 1 - 2\epsilon$$

if and only if the spectrum of pqp in pAp is contained in $(\epsilon, 1 - \epsilon)$.

PROOF. Write $a = pqp$, and consider the abelian C^*–algebra generated by a and p, which we can take to be the identity. In this notation we seek to show

$$\|2a - 1\| = \max\{2\|a\| - 1,\ 1 - 2\|1 - a\|\}.$$

Write α (resp. β) for the maximum (resp. minimum) element in the spectrum of a, and note that $\alpha = \|a\|$, $1 - \beta = \|1 - a\|$. We have

$$\|2a - 1\| = \begin{cases} 2\alpha - 1 & \text{if } \alpha + \beta \geq 1 \\ 1 - 2\beta & \text{if } \alpha + \beta \leq 1, \end{cases}$$

as asserted. Finally,

$$\|p(2q - 1)p\| < 1 - 2\epsilon$$

holds if and only if

$$2\alpha - 1 < 1 - 2\epsilon \ \text{ and } \ 1 - 2\beta < 1 - 2\epsilon,$$

and these conditions are true if and only if

$$\alpha < 1 - \epsilon \ \text{ and } \ \beta > \epsilon.$$

□

PROPOSITION 7.4. *Suppose H has finite dimension, p is a projection in $B(H)$ of rank r, q is a projection in $\mathcal{D}$ and $0 < \epsilon < 1/2$. The following are equivalent.*

  i) $\|qpq + (1 - q)p(1 - q)\| < 1 - \epsilon$.
  ii) *qpq has rank r and*

$$\left\|qpq - \frac{1}{2}e\right\| < \frac{1}{2} - \epsilon,$$

  where e denotes the projection onto the range of qpq.
  iii) *If $s = 2q - 1$, then $\|\Phi(s)\| = \|psp\| < 1 - 2\epsilon$.*

PROOF. Assume that i) holds. We have then that

$$\|qpq\| < 1 \quad \text{and} \quad \|(1 - q)p(1 - q)\| < 1.$$

We claim that qpq has rank r. Indeed, if not, the rank of qpq is strictly less than r, and so intersection of the null space of qpq and the range of p contains a unit vector η. But then

$$\|pq\eta\|^2 = (pq\eta, pq\eta) = (qpq\eta, \eta) = 0,$$

so that $\eta = p\eta = p(1 - q)\eta$. In this case we would have

$$1 = \|p(1 - q)\eta\|^2 = ((1 - q)p(1 - q)\eta, \eta),$$

and by Cauchy–Schwarz, $(1 - q)p(1 - q)\eta = \eta$. Since $\|(1 - q)p(1 - q)\| < 1$, this is impossible, and so qpq has rank r.

Thus, Proposition 7.2 applies (with $b = qpq$ and $c = (1 - q)p(1 - q)$). Using the notation of that proposition we have

$$\|e - b\| = \|c\| < 1 - \epsilon,$$

and so the smallest positive eigenvalue of qpq is $> \epsilon$. Hence ii) is true.

Next, if ii) is true, then the positive eigenvalues of qpq lie between ϵ and $1 - \epsilon$ and therefore by Proposition 7.2 the same is true for $(1 - q)p(1 - q)$. Thus, i) and ii) are equivalent.

Finally note that by Lemma 7.3, $\|\Phi(s)\| < 1 - 2\epsilon$ if and only if the spectrum of pqp (in $B(pH)$) is contained in the interval $(\epsilon, 1 - \epsilon)$, and this containment holds if and only if $\|pqp\| < 1 - \epsilon$ and $\|p(1 - q)p\| < 1 - \epsilon$. Since $\|qpq\| = \|pqp\|$ and $\|(1 - q)p(1 - q)\| = \|p(1 - q)p\|$, iii) and i) are equivalent. $\square$

With these preparations we are ready to discuss the connection between our present investigation and the problem raised by Kadison and Singer. Fix an orthonormal basis $\{\eta_j : j \in S\}$ for the Hilbert space H such that the elements of the MASA $\mathcal{D}$ are diagonal with respect to this basis. (Here the index set S may be finite or countably infinite). If q_j denotes the minimal projection onto $\text{span}\{\eta_j\}$, then we have that

$$\text{tr}(\Phi(q_j)) = \text{tr}(pq_jp) = \text{tr}(q_jpq_j) = (p\eta_j, \eta_j),$$

and so

$$\delta(p) = \sup\{(p\eta_j, \eta_j) : j \in S\}.$$

Thus, δ measures the size of the diagonal entries in the matrix for p given by the basis $\{\eta_j : j \in S\}$.

CONJECTURE 7.5. *Suppose H has finite dimension. There exist constants γ and ϵ with $0 < \gamma \leq 1/2$ and $0 < \epsilon < 1/2$ such that if p is a projection with the property that*

$$\delta = \sup\{(p\eta_j, \eta_j) : j \in S\} < \gamma,$$

then there is a projection q in $\mathcal{D}$ such that

$$\|qpq + (1 - q)p(1 - q)\| < 1 - \epsilon.$$

It has been known for some time [5] that if conjecture 7.5 is true then the problem of Kadison and Singer has a positive answer. (See [43] for an elegant reduction of the infinite–dimensional case to the finite–dimensional setting.) Hence, by Proposition 7.4, if Conjecture 7.1.3 is true then each pure state on $\mathcal{D}$ has a unique state extension to $B(H)$. For the sake of completeness we include a proof of this fact.

PROPOSITION 7.6. *The truth of Conjecture 7.1.3 implies the truth of conjecture 7.1.2.*

PROOF. Suppose that 7.1.3 holds and fix a projection p in $B(H)$. First suppose that p has finite rank. We have then that $q_\alpha p q_\alpha \to p$ in norm as q_α ranges over the directed family of finite rank projections in $\mathcal{D}$. By spectral theory this means that, if r_α denotes the range projection of $q_\alpha p q_\alpha$, then $\{r_\alpha\}$ also converges in norm to p. Note that for "large" α, we have that $|\delta(r_\alpha) - \delta(p)|$ is small. Hence, Conjecture 7.1.3 applies to the map $\Phi_\alpha(a) = r_\alpha a r_\alpha$; so, it follows that for each such α there exists a symmetry s_α in $\mathcal{D}$ such that $\|r_\alpha s_\alpha r_\alpha\| < 1 - 2\epsilon$. Since $\mathcal{D}$ is atomic and abelian, its symmetry group is weak* compact. Hence, there is a subnet $\{s_\beta\}$ that converges to a symmetry s in the weak operator topology. Since weak operator convergence is norm decreasing, we have

$$\|psp\| \le \limsup_\beta \|r_\beta s_\beta r_\beta\| \le 1 - 2\epsilon.$$

Hence, decreasing ϵ if necessary, conjecture 7.1.2 holds for projections of finite rank.

Now allow p to have infinite rank and let $\{p_\alpha\}$ denote a net of finite rank projections that converges up to p. We have that $\delta(p_\alpha) \le \delta(p)$, and so, by the previous paragraph, there exist symmetries $\{s_\alpha\}$ in $\mathcal{D}$ such that

$$\|p_\alpha s_\alpha s p_\alpha\| \le 1 - 2\epsilon.$$

Passing to a subnet as above, we may assume that $\{s_\alpha\}$ converges to a symmetry s in $\mathcal{D}$ and again conclude

$$\|psp\| \le \limsup_\alpha \|p_\alpha s_\alpha p_\alpha\| \le 1 - 2\epsilon.$$

Thus, conjecture 7.1.2 is true. □

PROPOSITION 7.7. *If Conjecture 7.1.2 is true, then each pure state h on $\mathcal{D}$ has a unique extension to a pure state on $B(H)$.*

PROOF. Let f denote the natural extension of h to $B(H)$ given by

$$f(x) = h(D(x)),$$

where D denotes the projection of a matrix onto its diagonal. Suppose that the Proposition is false so that f is not the unique state extension of h. In this case, the (closed) projection r in $\mathcal{D}^{**} \subset B(H)^{**}$ that supports h [42,p.134 and 37,3.13.6] is not minimal in $B(H)^{**}$. On the other hand, by [6], the state f is pure and so its support projection s is minimal by [37,3.13.6 and 3.2.3] and $s < r$. Hence, by [37,4.3.15], there is a pure state g such that g is orthogonal to f, $f(r) = g(r) = 1$ and $f|_\mathcal{D} = g|_\mathcal{D} = h$. If we write t for the support projection of g, then s and t are rank 1 closed projections. Since f and g are orthogonal, we have $st = 0$. By the noncommutative Urysohn Lemma [2,Theorem 1], there exists a positive norm 1 element c in $B(H)$ such that $cs = 0$ and $ct = 1$. It follows that $f(c) = 0$ and $g(c) = 1$. By [3, Proposition 2.3] we may find an excising net $\{q_\alpha\}$ of projections for h in $\mathcal{D}$. Thus, $h(q_\alpha) = 1$ for each α and

$$\|q_\alpha D(c)q_\alpha - h(D(c))\| = \|q_\alpha D(c)q_\alpha - f(c)q_\alpha\| = \|q_\alpha D(c)q_\alpha\| \to 0,$$

so we may set $b = q_\alpha c q_\alpha$ for suitable α and get

$$\|D(b)\| = \|D(q_\alpha c q_\alpha)\| = \|q_\alpha D(c)q_\alpha\| < \gamma/2.$$

(Here γ is the constant given in Conjecture 7.1.2). Write p for the spectral projection of b corresponding to the interval $[\frac{1}{2}, 1]$. We have $p < 2b$, so

$$\|D(p)\| < 2\|D(b)\| < \gamma.$$

Now apply Conjecture 7.1.2, using the projection p as above in the definition of the map Φ, to get a symmetry s in $\mathcal{D}$ such that $\|psp\| < 1$. On the other hand, $g(b) = 1$, so $g(p) = 1$; hence

$$|g(psp)| = |g(s)| = |h(s)| = 1,$$

since s is a symmetry in $\mathcal{D}$ and h is pure on $\mathcal{D}$. This contradicts $\|psp\| < 1$. Hence, no such g can exist and the proposition is proved. $\square$

The main result in this section (Theorem 7.12) asserts that Conjecture 7.1.2 is true when the projection p is of finite rank. This is an example of a Lyapunov theorem of Type 4. The proof of the theorem relies on a result from the theory of matroids due to Edmunds and Fulkerson [19]. In order to state this theorem, it is necessary to introduce some definitions and notation.

The notion of a matroid was invented to study the properties of linear independence of vectors in an abstract setting. Informally, a matroid is an abstract collection of "linearly independent" sets. More precisely, a **matroid** is a pair $(S, \mathcal{I})$, where S is a finite set and $\mathcal{I}$ is a non–void family of subsets of S. The elements of $\mathcal{I}$ are called the **independent** sets and must satisfy the following properties:

(1) Every subset of an independent set is independent.
(2) Given a subset σ of S, all maximal independent subsets of σ have the same cardinality, called the **rank** of σ.

The rank of σ is denoted by $\rho(\sigma)$. The prototypical example of a matroid is given by a finite set S of elements in a vector space with the family of linearly independent subsets of S serving as $\mathcal{I}$. We will use the following version of the

Edmunds–Fulkerson theorem [19,Theorem 2b, p.150]:

> THEOREM. (EDMUNDS AND FULKERSON). *A matroid* $(S, \mathcal{I})$ *contains* m *mutually disjoint independent subsets of cardinality* r *if and only if for every subset* σ *of* S *we have*
>
> $(*)$ $\qquad\qquad\qquad |\sigma| \geq m(r - \min\{r, \rho(\sigma')\}),$
>
> *where* $|\sigma|$ *denotes the cardinality of* σ *and* $\sigma' = S \backslash \sigma$. *In case* $r = \rho(S)$, *then* $r \geq \rho(\sigma')$ *for every subset* σ *and so the condition* $(*)$ *reduces to the requirement that*
>
> $(**)$ $\qquad\qquad\qquad |\sigma| \geq m(r - \rho(\sigma'))$
>
> *for each subset* σ *of* S.

We shall study certain matroids that are determined by matrices. The following lemma provides the key step in establishing that our definition produces a matroid. Recall that if v and w are vectors in a Hilbert space H, then $v \otimes v$ denotes the operator on H defined by $(v \otimes v)(w) = (w, v)v$.

LEMMA 7.8. *If* $v_1, \ldots, v_r$ *are vectors in* H, *and if we write* $a_i = v_i \otimes v_i$, *for* $i = 1, \ldots, r$, *then the rank of* $a = a_1 + \cdots + a_r$ *is equal to the linear dimension of the span of* $\{v_1, \ldots, v_r\}$.

PROOF. Let s denote the rank of a and write t for the dimension of $V = \text{span}\{v_i : i = 1, \ldots, r\}$. Since a is 0 on the orthogonal complement of V, we have $s \leq t$.

For the reverse inequality, first suppose $v_1, \ldots, v_r$ are linearly independent so that $t = r$. Replacing H by the linear span of $\{v_1, \ldots, v_r\}$ if necessary, we may assume $n = t = r$. Fix a vector v in the null space of a. We have

$$av = \sum_{i=1}^{r} (v, v_i) v_i = 0.$$

Since the v_i's are linearly independent, $(v, v_i) = 0$ for $i = 1, \ldots, n$ and, since the v_i's span H, we get $v = 0$. Thus, the null space of a is $\{0\}$ and the matrix a has rank $n = r = t$.

Now suppose V has dimension $t < r$. If $t = 0$, the result is trivial; so, assume $0 < t < r$. Relabeling if necessary, we may assume $v_1, \ldots, v_t$ are linearly independent. Consider

$$b = \sum_{i=1}^{t} a_i.$$

By the first part of the proof, the rank of b is t. Since $0 \leq b \leq a$, the rank of a is at least t. Hence $s \geq t$ and the proof is complete. $\square$

DEFINITION 7.9. *Fix x in $M_n(\mathbb{C})$ and write $S = \{1, \ldots, n\}$. Define a family $\mathcal{I}(x)$ of subsets of S as follows. For each subset σ of S, let q_σ denote the associated diagonal projection in $M_n(\mathbb{C})$. Thus, the (i,i)–entry in q_σ is one if and only if $i \in \sigma$ and all other entries are 0. A subset σ is defined to be in $\mathcal{I}(x)$ if the matrix rank of $x^* q_\sigma x$ is equal to the cardinality $|\sigma|$ of σ. Also for each $\sigma \subset S$, let $\rho(\sigma)$ denote the rank of $x^* q_\sigma x$.*

PROPOSITION 7.10. *If x in $M_n(\mathbb{C})$, then $\{S, \mathcal{I}(x)\}$ is a matroid with rank function ρ as defined above.*

PROOF. Let $\{q_i\}$ denote the minimal diagonal projections in $M_n(\mathbb{C})$. For each i, $x^* q_i x$ is positive and of rank one. Hence, there is a vector v_i in $\mathbb{C}^n$ such that

$$x^* q_i x = v_i \otimes v_i.$$

Now fix a subset σ of S and let τ_1 and τ_2 denote maximal independent subsets of σ. If we write

$$a_j = x^* q_{\tau_j} x = \sum_{i \in \tau_j} v_i \otimes v_i$$

for $j = 1, 2$, then by Lemma 7.8,

$$\rho(\tau_j) = \text{the rank of } a_j = \text{the linear dimension of span}\{v_i : i \in \tau_j\}$$

for $j = 1, 2$. As τ_1 and τ_2 are maximal independent subsets of σ,

$$\text{span}\{v_i : i \in \tau_j\} = \text{span}\{v_i : i \in \sigma)\}$$

for $j = 1, 2$, and therefore $\rho(\tau_1) = \rho(\tau_2)$. Similarly, it follows from Lemma 7.8 that, if σ is an independent set and $\tau \subset \sigma$, then τ is independent. Hence $\{S, \mathcal{I}_x\}$ is a matroid with rank function $\rho(\cdot)$. □

THEOREM 7.11. *Suppose x is a normal n by n complex matrix such that the rank of x is r and $\|x\| \leq 1$. Let α_{ij} denote the (i,j)–entry of $x^* x$ and write*

$$\gamma = \max\{\alpha_{ii} : i = 1, \ldots, n\}.$$

If there is an integer $m > 1$ such that

$$\gamma \leq 1/m \text{ and } \operatorname{tr}(x^* x) > r - 1/m,$$

then there are mutually orthogonal diagonal projections $q_1, \ldots, q_m$ such that $\Sigma q_i = 1$ and $q_i x^* x q_i$ has rank r for $i = 1, \ldots, m$.

PROOF. We shall show condition $(**)$ in the Edmonds-Fulkerson theorem stated above holds for the matroid $(S, \mathcal{I}(x))$ defined in 7.9. Fix σ in $S = \{1, \ldots, n\}$ and let σ' denote the complement of σ in S. Since $\|x\| \leq 1$ and $\operatorname{tr}(x^* x) > r - 1/m$, we have

$$\rho(\sigma') \geq \operatorname{tr}(x^* q_{\sigma'} x) = \operatorname{tr}(x^* x) - \operatorname{tr}(x^* q_\sigma x) > r - \frac{1}{m} - \operatorname{tr}(x^* q_\sigma x).$$

Now

$$\operatorname{tr}(x^* q_\sigma x) = \operatorname{tr}(q_\sigma x x^* q_\sigma) \operatorname{tr}(q_\sigma x^* x q_\sigma) \le \gamma |\sigma| \le \frac{1}{m} |\sigma|,$$

since x is normal. Hence,

$$\rho(\sigma') > r - \frac{1}{m} - \frac{1}{m} |\sigma|.$$

Thus, for every subset σ of S,

$$r = \frac{1}{m} |\sigma| + r - \frac{1}{m} |\sigma| < \frac{1}{m} |\sigma| + \rho(\sigma') + \frac{1}{m}.$$

Therefore

$$|\sigma| > m \left(r - \rho(\sigma') - \frac{1}{m} \right) = m(r - \rho(\sigma')) - 1.$$

Since $|\sigma|$ and $m(r - \rho(\sigma'))$ are integers, we have

$$|\sigma| \ge m(r - \rho(\sigma')).$$

Thus, condition $(**)$ is satisfied and so, by the Edmunds–Fulkerson theorem, there are disjoint subsets $\sigma_1, \dots, \sigma_m$ of S such that $x^* q_{\sigma_i} x$ has rank r in $B(H)$ for $i = 1, \dots, m$. Therefore, $q_{\sigma_i} x x^* q_{\sigma_i} = q_{\sigma_i} x^* x q_{\sigma_i} = (x q_{\sigma_i})^* (x q_{\sigma_i})$ also has rank r for each i. Enlarging σ_1 if necessary, we may assume $\sum q_{\sigma_i} = 1$ and the proof is complete. $\square$

THEOREM 7.12. *Suppose m is an integer greater than one. If p is a finite rank projection in $B(H)$ such that $\delta(p) \le 1/m$, then there are mutually orthogonal projections $q_1, \dots, q_m$ in $\mathcal{D}$ such that $\Sigma q_i = 1$, and, if we write $s_i = 2q_i - 1$, then*

$$\|\Phi(s_i)\| = \|p s_i p\| < 1$$

for $i = 1, \dots, m$.

PROOF. Let $\{\eta_j\}$ denote an orthonormal basis for H that diagonalizes the elements of $\mathcal{D}$ and identify p with its matrix $\{\pi_{ij}\}$ in this basis. Since p has finite rank, which we denote by r, there is an integer n such that

$$(*) \qquad\qquad \sum_{i=1}^n \pi_{ii} > r - \frac{1}{m}.$$

Let a denote the n by n matrix with entries π_{ij}, $1 \le i, j \le n$ and set $b = \sqrt{a}$. We have that $\gamma = \max\{\pi_{ii} : 1 \le i \le n\} \le \delta(p) \le 1/m$ and $\operatorname{tr}(a) > r - 1/m$ by $(*)$ above. Applying Theorem 7.11 (with $x = b$), we get that there are mutually orthogonal diagonal projections $q_1, \dots, q_m$ that sum to 1_n (here 1_n denotes the $n \times n$ identity matrix) and such that $q_i a q_i$ has rank r for $i = 1, \dots, m$. Now replace q_1 by $q_1 + (1 - 1_n)$ so that we have have $\sum_{i=1}^m q_i = 1$ and note that the rank of $q_1 p q_1$ is unchanged. Further, since the matrix for $(1 - q_i) p (1 - q_i)$ with respect to the basis $\{\eta_j\}$ contains the matrix for $q_k p q_k$ for $k \neq i$, the rank of $(1 - q_i)p(1 - q_i)$ is r for $i = 1, \ldots, m$.

To complete the proof, it is enough to show that the spectrum of $pq_i p$ (considered as an element of $B(pH)$) is contained in $(0, 1)$, by Lemma 7.3. Since $q_i p q_i$ and $(1 - q_i)p(1 - q_i)$ each have rank r, so do $pq_i p$ and $p(1 - q_i)p$. Since p also has rank r, $pq_i p$ and $p(1 - q_i)p$ are invertible in $B(pH)$. Hence, the spectrum of $pq_i p$ is contained in $(0, 1]$. If $\|pq_i p\| = 1$, then there would be a unit vector η in pH such that $q_i \eta = \eta = p\eta$. But then, letting e denote the projection onto the orthogonal complement of $\{\eta\}$ in pH, we would have that the rank of pe is $r - 1$ and $p(1 - q_i)p \leq pe$, so that the rank of $p(1 - q_i)p$ would be less than r. Since this is not the case, $\|pq_i p\| < 1$ and the spectrum of $pq_i p$ lies in $(0, 1)$, as desired. $\square$

REMARKS. 1) Note that if the conclusion of Theorem 7.12 is true then (in the notation of the theorem) we have $q_1 + \cdots + q_m = 1$ and

$$\|(1 - q_i)p(1 - q_i)\| < 1$$

for $i = 1, \ldots, m$ by Proposition 7.4.

2) Suppose p is a projection of rank r acting on a Hilbert space H of dimension n. If $\{w_1, \ldots, w_r\}$ denotes an orthonormal basis for the range of p and $\{\eta_1, \ldots, \eta_n\}$ an orthonormal basis for H, then we may view each w_i as an 1 by n matrix whose entries ω_{ij} are coefficients of w_i with respect to the basis $\{\eta_j\}$. If we write

$$w = \begin{pmatrix} \omega_{11} & \cdots & \omega_{1n} \\ \vdots & \ddots & \vdots \\ \omega_{r1} & \cdots & \omega_{rn} \end{pmatrix}$$

and if we also identify p with its matrix determined by $\{\eta_j\}$, then we have $ww^* = 1_r$ (the r by r identity matrix) and $w^*w = p$. In fact, if we write $v_1, \ldots, v_n$ for the columns of w, then we have that the (i, j)–entry in p is $v_i^* v_j$. Hence, in this notation, the number δ defined above is given by

$$\delta = \max \left\{ \|v_j\|_2^2 \right\}.$$

Also, if we put $a_j = v_j v_j^*$ then, since the v_j's are orthonormal, we have

$$a_1 + \cdots + a_n = ww^* = 1_r.$$

Note that each a_j is positive and has rank 1. Moreover,

$$\|a_j\| = \operatorname{tr}(a_j) = \operatorname{tr}(v_j^* v_j) \leq \delta(p).$$

Given these observations it is not too hard to show that Conjecture 7.1.3 is equivalent to the following:

CONJECTURE 7.13. *There are constants γ and ϵ with $0 < \gamma \leq 1/2$ and $0 < \epsilon < 1/2$ such that, if $a_1, \ldots, a_n$ are positive r by r matrices of rank one satisfying*

   i) $a_1 + \cdots + a_n = 1_r$ *and*
   ii) $\max\{\|a_j\|\} = \delta < \gamma$,

then there is a subset σ of $\{1, \ldots, n\}$ such that, with

$$a_\sigma = \sum_{j \in \sigma} a_j,$$

we have

$$\|a_\sigma - (1/2)1_r\| < 1 - 2\epsilon.$$

This conjecture has been extensively tested on a computer with randomly generated projections p of rank as high as 30. No counterexample has been found. In fact, in every test case a subset σ was found such that

$$\|a_\sigma - (1/2)1_r\| \leq \delta.$$

Thus, it appears that—at least generically—the stronger result, with δ replacing $1 - 2\epsilon$ (which is then equivalent to Conjecture 7.1), may be true.

Finally, we note that Corollary 4.4 shows that a weak version of 7.13 is true. Indeed, if we apply Corollary 4.4 to the matrix

$$a = \begin{pmatrix} |\omega_{11}|^2 & \cdots & |\omega_{1n}|^2 \\ \vdots & \ddots & \vdots \\ |\omega_{r1}|^2 & \cdots & |\omega_{rn}|^2 \end{pmatrix},$$

then, since the rows of w are orthonormal and the (j, j)–entry of p is precisely the sum of the jth column of a, we have that

$$\gamma = \max_j \left\{ \sum_{i=1}^r |\omega_{ij}|^2 \right\} = \max_j \left\{ \|v_j\|_2^2 \right\} = \delta(p)$$

and the row sums of a are 1. Hence, if we write $v = \langle \frac{1}{2}, \ldots, \frac{1}{2} \rangle$ and apply Corollary 4.4, there is a subset σ of $\{1, \ldots, n\}$ such that

$$\left| \sum_{i \in \sigma} |\omega_{ij}|^2 - \frac{1}{2} \right| < \delta, \qquad j = 1, \ldots, n.$$

We conclude this section by presenting 2 examples. Note that the map $a \mapsto pap$ is both positive and of norm one on $\mathcal{D}$. In the first example we show that it is possible to find a positive (but not norm 1) map Ψ from $\mathcal{D}$ into $B(H)$ for which none of our conjectures hold.

EXAMPLE 7.14. Suppose H has finite dimension n and let $\{\eta_1, \ldots, \eta_n\}$ denote an orthonormal basis for H. Write p_j for the orthogonal projection on the (1–dimensional) span of $\{\eta_j + \eta_1\}$ for $j = 2, \ldots, n$, let $\{q_j\}$ denote the minimal projections in $\mathcal{D}$, and define $\Psi : \mathcal{D} \to B(H)$ by

$$\Psi\left(\sum_{j=1}^n \lambda_j q_j\right) = \sum_{j=2}^n \frac{2\lambda_j}{\sqrt{n-1}} p_j.$$

For any symmetry $s = \Sigma \epsilon_j q_j$ we have

$$\|\Psi(s)\eta_1\|^2 = \left\|\sum_{j=2}^{n} \frac{\epsilon_j}{\sqrt{n-1}}(\eta_1 + \eta_j)\right\|^2$$

$$> \frac{1}{n-1}\left\|\sum_{j=2}^{n} \epsilon_j \eta_j\right\|^2 = 1.$$

However, the value of $\delta = \delta(\Psi)$ in this case is $2(n-1)^{-1/2}$, which can be made as small as we like.

In our final example we show that Conjecture 7.1.2 is false for $\delta = \gamma = 1/2$. Specifically, for any $\epsilon > 0$ there exists a projection p on a subspace of H with the diagonal of p less than $1/2$ such that $\|pup\| > 1 - 2\epsilon$ for every diagonal unitary operator u. By taking direct sums we get a projection p' such that $\delta(p') = 1/2$ and $\|p'up'\| = 1$ for every diagonal unitary u.

EXAMPLE 7.15. Fix $0 < \epsilon < 1/2$ and select $\rho > 0$ such that $\rho^2 < \epsilon/2$. Next pick a positive integer m so that, with $s = 2^m$, we have

$$(*) \qquad\qquad (1 - \rho^2)\left(\frac{1}{2} + \frac{1}{s}\right) < \frac{1}{2}.$$

Since s is a power of 2, there is an orthonormal basis $\{v_j\}$ of $\mathbb{R}^s$ such that, if $\{\eta_j\}$ denotes the standard basis for $\mathbb{R}^s$, then

$$(v_j, \eta_k) = \pm \frac{1}{\sqrt{s}}$$

for all j, k between 0 and s. For example, such a basis may be obtained from the rows of a Hadamard matrix. Alternatively, $\{v_j\}$ may be constructed directly from tensor products of the 2–dimensional vectors

$$\frac{1}{\sqrt{2}}(1,1) \qquad \text{and} \qquad \frac{1}{\sqrt{2}}(1,-1).$$

Now set $r = s/2 + 1$ and $n = s + r$. View each v_j as a vector in $H = \mathbb{C}^n$ by setting the undefined coordinates to 0. For $1 \leq j \leq r$ write

$$w_j = \rho' v_j + \rho \eta_{s+j},$$

where $\rho' = \sqrt{1 - \rho^2}$. Let W denote the linear span of $\{w_1, \ldots, w_r\}$ and write p for the projection of H onto W. If we write V for the subspace spanned by $\{v_1, \ldots, v_r\}$, then it is not difficult to verify that

$$(**) \qquad\qquad \|p - p_0\| < 2\rho^2,$$

where p_0 denotes the projection of H onto V.

Now fix a diagonal unitary u and let Y denote the linear span of $\{u^*x : x \in V\}$. Since u is diagonal, Y is contained in the span of $\{\eta_1, \ldots, \eta_s\}$. Since V is also contained in this subspace, and the sum of their dimensions exceeds s, there is a unit vector η in $Y \cap V$. As $\eta \in Y$, there is a unit vector ξ in V such that $\eta = u^*\xi$. Hence

$$(p_0 u p_0)\eta = (p_0 u)\eta = p_0(u\eta) = p_0(u(u^*\xi)) = p_0\xi = \xi = u\eta,$$

and so $\|p_0 u p_0\| = 1$. Thus, by (**) $\|pup\| > 1 - 4\rho^2 > 1 - 2\epsilon$.

Recall that $\delta = \delta(p)$ is the maximum of the diagonal entries in the matrix for p determined by the standard basis $\{\eta_j\}$. These entries are given as follows

$$(p\eta_j, \eta_j) = \sum_{i=1}^{r} |(\eta_j, w_i)|^2.$$

If $j > s$, then $(p\eta_j, \eta_j) = \rho^2$. On the other hand, if $j \leq s$ then

$$(p\eta_j, \eta_j) = \frac{r(1 - \rho^2)}{s} = (1 - \rho^2)\left(\frac{1}{2} + \frac{1}{s}\right).$$

Using (*) above and the fact that $\frac{r}{s} > \frac{1}{2}$, we get

$$\frac{1 - \rho^2}{2} < \delta(p) < \frac{1}{2}.$$

Hence, p has the desired properties.

§8. AN APPLICATION TO THE PAVING PROBLEM

Throughout this section we use the notation of Section 7. Thus $\mathcal{D}$ stands for an atomic NASA in $B(H)$, where H is now assumed to be of infinite dimension, and we write D for the projection of $B(H)$ onto $\mathcal{D}$. Fix an infinite set $\{q_j\}$ of orthogonal finite rank projections in $\mathcal{D}$ with sum 1 and let $\mathcal{M}$ stand for the commutant of $\{q_j\}$ in $B(H)$. Recall that we say that an element x in $B(H)$ can be **paved** if, for every $\epsilon > 0$, there is a finite family $\{t_i\}$ of projections in $\mathcal{D}$ with sum 1 such that

$$(*) \qquad\qquad \|t_i(x - D(x))t_i\| < \epsilon.$$

for each i. It is well–known [5] that, if the dimension of the direct summands in $\mathcal{M}$ is unbounded, then the problem of Kadison and Singer discussed in Section 7 has a positive answer if and only if every projection in $\mathcal{M}$ can be paved. Note that it suffices to find a finite set $\{t_i : i = 1, \dots, m\}$ satisfying $(*)$ and such that $1 - (t_1 + \cdots + t_m)$ has finite rank, since in that case the family $\{t_i'\}$ formed by taking the union of the t_i's with the minimal projections in $\mathcal{D}$ that are orthogonal to the t_i's has the desired property. Also, if the rank of the q_j's is bounded by r, then we may find projections $t_1, \dots, t_r$ in $\mathcal{D}$ with sum 1 and such that $t_i q_j$ has rank 0 or 1 for each i and j. In this case for any x in $\mathcal{M}$ we have

$$t_i x t_i = D(x)t_i$$

and therefore x can be paved.

We begin by recording some familiar facts that will be used in the proof of the main result of this section.

LEMMA 8.1. *If $a = \{\alpha_{ij}\}$ is an r by r complex matrix then*

i) $\|a\|_1 = \sum_{i,j=1}^{n} |\alpha_{ij}| \leq r\sqrt{\sum_{i,j=1}^{n} |\alpha_{ij}|^2} = r\|a\|_2$ *and*

ii) $\|a\| \leq r \max\{|\alpha_{ij}|\} = r\|a\|_\infty,$

where $\|\cdot\|$ denotes the operator norm and the other norms are as given above.

PROOF. We have

$$\left(\sum |\alpha_{ij}|\right)^2 \leq r^2 \sum |\alpha_{ij}|^2$$

by the Schwarz inequality, and so i) is true. For ii), fix a unit vector η in $\mathbb{C}^r$. If we write γ_j for the coefficients of η, then, using the Schwarz inequality once more, we have

$$\sum_{j=1}^{r} |\alpha_{ij}\gamma_j| \leq \sqrt{r}\|a\|_\infty$$

for each i. Hence,

$$\|a\eta\| \le r\|a\|_\infty,$$

as desired. ☐

Fix a projection p in $\mathcal{M}$. For each j, write $p_j = pq_j$ and define $\delta_j = \delta_j(p)$ by the formula

$$\delta_j = \|D(p_j)\|.$$

Let r_j denote the rank of p_j.

THEOREM 8.2. *If* $\lim_{j\to\infty} \delta_j(r_j)^2 = 0$, *then* p *can be paved.*

PROOF. Fix a positive integer j and let $\{\eta_k : k = 1, \dots, r_j\}$ denote an orthonormal basis for $p_j H$. Write Ψ_j for the map from $\mathcal{D}$ to $\ell^\infty(r_j{}^2)$ defined by

$$\Psi_j(x) = \langle (x\eta_k, \eta_l) \rangle.$$

Thus, we have

$$\|\Psi_j(x)\|_\infty = \sup\{|(x\eta_k, \eta_l)| : 1 \le k, l \le r_j\}.$$

As $\Psi_j(x)$ is also an $r_j \times r_j$ matrix, it has an operator norm, which we denote by $\|\psi_j(x)\|_{op}$. If q is a minimal projection in $\mathcal{D}$, then by i) of Lemma 8.1 we have

$$\|\Psi_j(q)\|_1 \le r_j \|p_j q p_j\|_2 = r_j \sqrt{\operatorname{tr}(p_j q p_j q p_j)}$$

$$= r_j \sqrt{\operatorname{tr}((q p_j q)^2)} = r_j \operatorname{tr}(q p_j q) \le r_j \delta_j,$$

where we have used the fact that the trace is multiplicative on $q p_j q$ because its rank is one. Thus, in the notation of Sections 4 and 7, we have $\delta(\Psi_j) \le r_j \delta_j$. Now fix $0 < \epsilon < 1$ and select an integer j_0 such that, if $j > j_0$, then $\delta_j r_j{}^2 < \epsilon$. Also, pick an integer m satisfying

$$\frac{1}{4\epsilon} < m < \frac{1}{3\epsilon} < \frac{1}{3\delta_j r_j}$$

for each $j > j_0$. By Theorem 5.7 there are diagonal projections

$$\{q_{ij} : i = 1, \dots, m\}$$

that sum to q_j such that

$$\|\Psi_j(q_{ij} - (1/m)q_j)\|_\infty < 3\delta_j r_j.$$

Since $m < 1/3\delta_j r_j$, each $q_{ij} \ne 0$ by remark 2) following the proof of Theorem 5.7. By part ii) of Lemma 8.1 we have

$$\|\Psi_j(q_{ij} - (1/m)q_j)\|_{op} < 3\delta_j r_j{}^2.$$

Also,
$$\|\Psi_j(q_j)\|_{op} = 1,$$
because $\Psi_j(q_j)$ is the identity matrix. Hence,
$$\|p_j q_{ij} p_j\| = \|\Psi_j(q_{ij})\|_{op} < \frac{1}{m} + 3\delta_j r_j^2 < 4\epsilon + 3\epsilon = 7\epsilon.$$

Since $\|p_j q_{ij} p_j\| = \|q_{ij} p_j q_{ij}\|$ and $\|D(p_j)\| = \delta_j < \epsilon$, we get
$$\|q_{ij}(p_j - D(p_j))q_{ij}\| = \|q_{ij}(p - D(p))q_{ij}\| < 8\epsilon.$$

If we take
$$t_i = \sum_{j>j_0} q_{ij}$$
for $i = 1, \ldots, m$, then $\|t_i(p - D(p)t_i\| < 8\epsilon$ for each i. Since $1 - (t_1 + \cdots + t_m) = q_1 + \cdots + q_{j_0}$ is finite–dimensional, it follows, as noted in the remark in the first paragraph of this section, that p can be paved. $\square$

Note that in order to pave an element of $\mathcal{M}$, we need only find a finite family of projections $\{t_i\}$ such that $\|t_i(x - D(x))t_i\|$ is small. In fact, it is sufficient to show that if $\|x\| \leq 1$, then we can we can always find $\{t_i : i = 1, \ldots, m\}$ such that

(∗) $\|t_i(x - D(x))t_i\| < 1 \quad i = 1, \ldots, m.$

Indeed, in this case, by taking direct sums, it follows that there is $0 < \epsilon < 1/2$, independent of x, such that (∗) holds with $1 - \epsilon$ replacing 1. Hence, in this case, for each positive integer k, we may find diagonal projections $\{t_i : i = 1, \ldots, m^k\}$ such that
$$\|t_i(x - D(x))t_i\| < (1 - \epsilon)^k \quad i = 1, \ldots, m^k.$$

The question arises then as to how many t_i's are necessary to yield (∗). Given $x \in \mathcal{M}$, let us write
$$\alpha_m(x) = \inf\{\| \textstyle\sum_{i=1}^{m} t_i(x - D(x))t_i\| : t_1 \in \mathcal{D}, t_i = t_i^2 \text{ and } \textstyle\sum_{i=1}^{m} t_i = 1\}.$$

In the following result we continue to use the notation developed above.

COROLLARY 8.3. If $\delta_j < 1/2$ for all j, and if $\limsup_{j\to\infty} \delta_j r_j^2 < 1/2$, then $\alpha_2(p) < 1$.

PROOF. Since $\limsup_{j\to\infty} \delta_j r_j^2 < 1/2$, we may fix m and γ so that, for $j > m$ we have $|\delta_j r_j^2| < \gamma < 1/2$. Since $\delta_j < 1/2$, by 7.12 there are symmetries s_j such that $\|p_j s_j p_j\| < 1$, $j = 1, ..., m$. For $j > m$ we define Ψ_j as in the proof of 8.2. As in the proof of Theorem 8.2, we have $\delta(\Psi_j) < r_j \delta_j$, and so by Theorem 4.3(ii) we may find symmetries s_j $(j > m)$ such that $\|\Psi_j(s_j)\|_\infty < 2\delta(\Psi_j)$. By Lemma 8.1(ii) for $j > m$ we have
$$\|\Psi_j(s_j)\| < r_j \|\Psi_j(s_j)\|_\infty < 2\delta(\Psi_j)r_j < 2\delta_j r_j^2 < 2\gamma,$$

where we used the fact that for $j > m$, $\delta_j r_j^2 < \gamma < 1/2$. If we put $s = \sum_{j=1}^{\infty} s_j$, then s is a symmetry and $\|psp\| < 1$. We have $s = 2q - 1$ for a diagonal projection q. By Prop. 7.4, $\|qpq + (1-q)p(1-q)\| < 1$, and one can check that $\|q(p - D(p))q + (1-q)(p - D(p))(1-q)| < 1$ because $\|D(p)\| < 1/2 < 1$. $\square$

Remark:. In a recent paper [13] Bourgain and Tzafriri use much more complicated methods and get a stronger result than Theorem 8.2, but their methods do not directly yield Corllary 8.3. They show that if there is $\gamma > 0$ such that

$$\delta_j \leq \frac{1}{(\log r_j)^{1+\gamma}},$$

then p can be paved.

REFERENCES

1. C. A. Akemann, *The dual space of an operator algebra*, Trans. Amer. Math. Soc. **120** (1967), 286–302.
2. C. A. Akemann, *Left ideal structure of C*-algebras*, J. Functional Analysis **6** (1970), 305–317.
3. C. A. Akemann, J. Anderson and Gert K. Pedersen, *Excising states in C*-algebras*, Canad. J. Math. **38** (1986), 1239–1260.
4. C. A. Akemann, Gert K. Pedersen, *Facial Structure in Operator Algebra Theory*, To appear, J. London Math. Soc..
5. J. Anderson, *A conjecture concerning the pure states of B(H) and a related theorem*, (5th International Conference on Operator Theory), Timisoara 1980 Operator Theory 2, Birkhauser Verlag, 1981, pp. 25–43.
6. J. Anderson, *Extreme points in sets of positive linear maps on B(H)*, J. Functional Analysis **31** (1979), 195–217.
7. T. E. Armstrong and K. Prikery, *Liapounoff's theorem for nonatomic finitely additive, bounded, finite dimensional, vector-valued measures*, Trans. Amer. Math. Soc. **266** (1981), 499–514.
8. Z. Arstein, *Yet another proof of the Lyapunov convexity theorem*, Proc. Amer. Math. Soc. **108** (1990), 89–91.
9. N. Azarnia and J. D. M. Wright, *On the Lyapunoff–Knowles theorem*, Quart. J. Math. Oxford (2) **33** (1982), 257–261.
10. J. Beck and Tibor Fiala, *Integer–making Theorems*, Discrete Appl. Math. **3** (1981), 1–8.
11. B. Blackadar, *Notes on the structure of projections in simple C*-algebras*, SemesterBericht Funktionanalysis, W82, Tubingen, March 1983.
12. _____ , *K–Theory for Operator Algebras*, Springer–Verlag, New York, 1986.
13. Bourgin and Tzafriri, *On a Problem of Kadison and Singer*, preprint.
14. L. G. Brown and G. K. Pedersen, *C*-algebras of real rank zero and their multiplier algebras*, preprint.
15. H. Choda, M. Enomoto, and M. Fujii, *Non commutative Liapounoff's theorem*, Math. Japonica **28 (5)** (1983), 651–653.
16. L. Dubins and E. Spanier, *How to cut a cake fairly*, Amer. Math. Monthly 68(1961), 1-17.
17. A. Dvoretzky, A. Wald, and J. Wolfowitz, *Relations among certain ranges of vector measures*, Pacific J. Math. 1 (1951), 59-74..
18. H. Dye, *The unitary structure in finite rings of operators*, Duke Math. J. **20** (1953), 55–69.
19. J. Edmunds and D. R. Fulkerson, *Transversals and Matroid Partition*, J. of Research of the National Bureau of Standards—B. Mathematics and Mathematical Physics **69B** (1965), 147–153.
20. C. M. Edwards and G. T. Ruttimann, *On the facial structure of a JBW* triple and its predual*, J. London Math. Soc. **38** (1986), 15–32.
21. R. E. Edwards, *Functional Analysis*, Holt, Rinehart and Winston, New York, 1965.
22. E. Effros, *Order ideals in a C*-algebra and its dual*, Duke Math. J. **33** (1963), 391–411.
23. J. Elton and T. P. Hill, *A generalization of Lyapounov's convexity theorem to measures with atoms*, Proc. Amer. Math. Soc. **99** (1987), 297–304.

24. I. Glicksberg, *An analogue of Liapounoff's convexity theorem for Birnbaum–Orlicz spaces and the extreme points of their unit balls*, Pacific J. Math. **116** (1985), 265.

25. P. R. Halmos, *Normal dialations and extensions of operators*, Summa Brasilensis Mathematicae **2 (1950), 125-134**.

26. T. Jech, *Set Theory*, Pure and applied mathematics, vol. 79, Academic Press, New York, 1978.

27. R. V. Kadison and J. R. Ringrose, *Fundamentals of the Theory of Operator Algebras I (II)*, Academic Press, New York, 1983 (1986).

28. R. V. Kadison and I. M. Singer, *Extensions of pure states*, Amer J. Math **81** (1959).

29. J. Kingman and A. Robertson, *On a theorem of Lyapunov*, J. London Math. Soc. **43** (1968), 347–351.

30. J. Legut, *Inequalities for α-optimal partitioning of a measurable space*, Proc. Amer. Math. Soc. 104 (1988), 1249-1251.

31. J. Lindenstrauss, *A short proof of Liapanov's convexity theorem*, J. Math. and Mech. **15** (1966), 971–972.

32. P. Loeb, *A combinatorial analog of Lyapunov's theorem for infinitesimally generated atomic vector measures*, Proc. Amer. Math. Soc. **39** (1973), 585–586.

33. A. Lyapunov, *On completely additive vector functions*, Bull. Acad. Sci. USSR **4** (1940), 465–478. (Russian)

34. A. Lyapunov, *On completely additive vector functions*, Izv. Akad. Nauk. SSSR **4** (1946), 277-279. (Russian)

35. A. W. Marshall and I. Olkin, *Inequalities, theory of majorization and it application*, Mathematics in science and engineering, vol. 143, Academic Press, New York, 1979.

36. C. Olech, *On the range of an unbounded vector–valued measure*, Math. Systems Theory **2** (1968), 251–256.

37. G. K. Pedersen, *C*-algebras and their Automorphism Groups*, Academic Press, London, 1979.

38. R. T. Rockafellar, *Convex Analysis*, Princeton University Press, Princeton, 1970.

39. W. Rudin, *Functional Analysis*, McGraw Hill, New York, 1973.

40. S. Sakai, *C*-algebras and W*-algebras*, Springer–Verlag, Berlin, 1971.

41. J. Spencer, *Six standard deviations suffice*, Trans. Amer. Math. Soc. **289** (1985), 679–706.

42. M. Takesaki, *Theory of Operator Algebras I*, Springer–Verlag, New York, 1979.

43. B. Tanbay, *Pure State Extensions and Compressibility of the ℓ_1-algebra*, preprint.

44. F. Tardella, *A new proof of the Lyapunov convexity theorem*, SIAM J. Control and Optimization **28 (2)** (1990), 478–481.

45. S. Tsui and S. Wright, *Asymptotic commutants and zeros of von Neumann algebras*, Math. Scand. **51 (2)** (1982), 232–240.

46. J. J. Uhl, *The range of a vector–valued measure*, Proc. Amer. Math. Soc. **23** (1969), 158–163.

47. D. Wulbert, *Lyapunov's and related theorems*, Proc. Amer. Math. Soc. **108 (4)** (1990), 955–960.

MEMOIRS of the American Mathematical Society

SUBMISSION. This journal is designed particularly for long research papers (and groups of cognate papers) in pure and applied mathematics. The papers, in general, are longer than those in the TRANSACTIONS of the American Mathematical Society, with which it shares an editorial committee. Mathematical papers intended for publication in the Memoirs should be addressed to one of the editors:

Ordinary differential equations, partial differential equations and applied mathematics to ROGER D. NUSSBAUM, Department of Mathematics, Rutgers University, New Brunswick, NJ 08903

Harmonic analysis, representation theory and Lie theory to AVNER D. ASH, Department of Mathematics, The Ohio State University, 231 West 18th Avenue, Columbus, OH 43210

Abstract analysis to MASAMICHI TAKESAKI, Department of Mathematics, University of California, Los Angeles, CA 90024

Real and harmonic analysis to DAVID JERISON, Department of Mathematics, M.I.T., Rm 2–180, Cambridge, MA 02139

Algebra and algebraic geometry to JUDITH D. SALLY, Department of Mathematics, Northwestern University, Evanston, IL 60208

Geometric topology and general topology to JAMES W. CANNON, Department of Mathematics, Brigham Young University, Provo, UT 84602

Algebraic topology and differential topology to RALPH COHEN, Department of Mathematics, Stanford University, Stanford, CA 94305

Global analysis and differential geometry to JERRY L. KAZDAN, Department of Mathematics, University of Pennsylvania, E1, Philadelphia, PA 19104-6395

Probability and statistics to RICHARD DURRETT, Department of Mathematics, Cornell University, Ithaca, NY 14853-7901

Combinatorics and number theory to CARL POMERANCE, Department of Mathematics, University of Georgia, Athens, GA 30602

Logic, set theory, general topology and universal algebra to JAMES E. BAUMGARTNER, Department of Mathematics, Dartmouth College, Hanover, NH 03755

Algebraic number theory, analytic number theory and modular forms to AUDREY TERRAS, Department of Mathematics, University of California at San Diego, La Jolla, CA 92093

Complex analysis and nonlinear partial differential equations to SUN-YUNG A. CHANG, Department of Mathematics, University of California at Los Angeles, Los Angeles, CA 90024

All other communications to the editors should be addressed to the Managing Editor, DAVID J. SALTMAN, Department of Mathematics, University of Texas at Austin, Austin, TX 78713.

General instructions to authors for

PREPARING REPRODUCTION COPY FOR MEMOIRS

> **For more detailed instructions send for AMS booklet, "A Guide for Authors of Memoirs."**
> **Write to Editorial Offices, American Mathematical Society, P.O. Box 6248,**
> **Providence, R.I. 02940-6248.**

MEMOIRS are printed by photo-offset from camera copy fully prepared by the author. This means that the finished book will look exactly like the copy submitted. Thus the author will want to use a good quality typewriter with a new, medium-inked black ribbon, and submit clean copy on the appropriate model paper.

Model Paper, provided at no cost by the AMS, is paper marked with blue lines that confine the copy to the appropriate size.

Special Characters may be filled in carefully freehand, using dense black ink, or **INSTANT** ("rub-on") **LETTERING** may be used. These may be available at a local art supply store.

Diagrams may be drawn in black ink either directly on the model sheet, or on a separate sheet and pasted with rubber cement into spaces left for them in the text. Ballpoint pen is not acceptable.

Page Headings (Running Heads) should be centered, in CAPITAL LETTERS (preferably), at the top of the page — just above the blue line and touching it.

LEFT-hand, EVEN-numbered pages should be headed with the AUTHOR'S NAME;

RIGHT-hand, ODD-numbered pages should be headed with the TITLE of the paper (in shortened form if necessary).

Exceptions: PAGE 1 and any other page that carries a display title require NO RUNNING HEADS.

Page Numbers should be at the top of the page, on the same line with the running heads.

LEFT-hand, EVEN numbers — flush with left margin;

RIGHT-hand, ODD numbers — flush with right margin.

Exceptions: PAGE 1 and any other page that carries a display title should have page number, centered below the text, on blue line provided.

FRONT MATTER PAGES should be numbered with Roman numerals (lower case), positioned below text in same manner as described above.

MEMOIRS FORMAT

> **It is suggested that the material be arranged in pages as indicated below.**
> **Note: Starred items (*) are requirements of publication.**

Front Matter (first pages in book, preceding main body of text).

Page i — *Title, *Author's name.

Page iii — Table of contents.

Page iv — *Abstract (at least 1 sentence and at most 300 words).

Key words and phrases, if desired. (A list which covers the content of the paper adequately enough to be useful for an information retrieval system.)

*1991 Mathematics Subject Classification. This classification represents the primary and secondary subjects of the paper, and the scheme can be found in Annual Subject Indexes of MATHEMATICAL REVIEWS beginnning in 1990.

Page 1 — Preface, introduction, or any other matter not belonging in body of text.

Footnotes: *Received by the editor date.
Support information — grants, credits, etc.

First Page Following Introduction – Chapter Title (dropped 1 inch from top line, and centered). Beginning of Text.

Last Page (at bottom) – Author's affiliation.